Stresses in glaciers

Peter Halfar

Stresses in glaciers

Methods of Calculation

Peter Halfar
Hamburg, Germany

ISBN 978-3-662-66023-2 ISBN 978-3-662-66024-9 (eBook)
https://doi.org/10.1007/978-3-662-66024-9

This book is a translation of the original German edition „Spannungen in Gletschern“ by Halfar, Peter, published by Springer-Verlag GmbH, DE in 2016. The translation was done with the help of artificial intelligence (machine translation by the service DeepL.com). A subsequent human revision was done primarily in terms of content, so that the book will read stylistically differently from a conventional translation. Springer Nature works continuously to further the development of tools for the production of books and on the related technologies to support the authors.

This Springer imprint is published by the registered company Springer-Verlag GmbH, DE, part of Springer Nature.
The registered company address is: Heidelberger Platz 3, 14197 Berlin, Germany

For Dorothea, Harry, Ronnie and Annelie

Preface to the German Edition

One would think that everything has already been said about a classic topic such as "Stresses in glaciers". But while studying this topic, I gradually got a different impression, because the simplicity of the balance conditions, which the forces and torques in glaciers have to satisfy, kept catching my eye. This simplicity of the balance conditions – they consist of the symmetry condition for the stress tensor and of a differential equation in which the divergence of the stress tensor occurs – led me to the assumption that there must also be a simple general solution. In fact, as expected, I found this general solution in a handbook on solid mechanics. The remarkable thing about this general solution was that it has never been applied to glacier dynamics. This gap shall be closed in the following.

After constructing this general solution of the balance conditions, I go one step further and compute the general solution that takes into account not only these balance conditions but also the boundary conditions at free glacier surfaces and at contact surfaces with standing water.

However, one cannot go further than the general solution of these balance and boundary conditions in general stress calculations if the conditions considered are to remain reliable and the computational effort justifiable. If one wanted to take into account the insufficiently known boundary conditions at the glacier bed or at the boundary surfaces to the neighbouring, not considered part of the glacier, the reliability would be lost. If the flow conditions – incompressibility of the ice and flow law – were to be taken into account, this would result in a high computational cost.

Therefore, with this general solution of the balance and boundary conditions, exactly that part of the stress calculations has been mastered, which is based on reliable conditions and can be carried out with reasonable computational effort. Because of its reliability, this general solution forms a solid starting point for all further calculations.

I have tried to explain the concept of the stress tensor in such detail and to present all calculation methods in such detail that a handbook and textbook for the calculation of stresses is available which is as self-contained as possible and which can be used without special knowledge and without further literature studies. A prerequisite for reading is a basic knowledge of analysis, distribution theory, linear algebra and classical mechanics. To enable the reader to quickly find and use what interests them, I have given an overview of the general solution as well as its possible applications in Chap. 9. I have also included numerous cross-references by footnotes to enable selective reading. In this way, and with the help of the table of contents, the reader can make sense of the book. Therefore, I have omitted an index.

The many formulas caused quite a problem. These formulas, which are supposed to contribute to the understanding of the presented calculation methods, could have the opposite effect if they interspersed the text too much and thus disturbed it too much. I tried to solve this problem by moving calculations to the appendix, but this was not enough. I found the solution to this problem only after I realized that there are two languages being spoken here, namely the language of the text and the language of the formulas, and that each language has its own message and that the more you mix the two languages, the more these messages interfere with each other. Therefore, I have separated the text and the formulas in each section and have avoided talking directly about the formulas in the text. Thus, in each section, the text and the formulas stand independently next to each other and are only loosely connected. This loose connection is established by the bracketed formula numbers interspersed in the text. These formula numbers are superscripted to indicate that they are not part of the text. In this way, the text remains undisturbed by the formulas and retains its independent meaning.

In order to also make the parts consisting only of formulas as understandable as possible, these parts are structured. For this reason, signs of equality are also occasionally – to specify them more exactly – provided with a hint, if an equality is not a condition but an equality according to a precondition, an equality by definition or a mathematical identity, or if another formula is to be considered. The special integration rules introduced in Chap. 3 and their symbolization by integral operators play an important role. With this not only many formulas can be written clearly but many calculations can also be done easily, because these integral operators satisfy especially simple rules of calculation. Thereby, not only ordinary functions but also distributions occur.

An explanation and a list of the symbols can be found at the end of the book. Only the obliquely crossed-out symbols are mentioned here, since they are not in general use. $\not{\mathbf{u}}$ denotes the antisymmetric tensor associated with a vector $\mathbf{u}$, and $\not{\mathbf{H}}$ denotes the vector associated with the antisymmetric part of a tensor $\mathbf{H}$.

Fortunate coincidences, for which I owe thanks to many people, have made it possible for me to write this book. I am especially grateful to my wife Dorothea. She was my first editor, and in dialogue with her, I have shaped the book.

Hamburg, Germany Peter Halfar
Summer 2015

Preface to the English Edition

It was very pleasant for me that Springer-Verlag decided to translate this book into English. In this translation, abstracts were added to every chapter. I have eliminated some misprints and inconsistencies, which occurred in the German edition.

I express my sincere gratitude to all who took part in this project.

Hamburg, Germany
2022

Peter Halfar

Contents

Part I

Introduction and Basics

Introduction

1

Abstract

The reason for the study is a long known mathematical method that is suitable for calculating stresses in glaciers, but which has never been applied in glaciology. The physical mechanisms are outlined. The concept of the investigation is to develop computational procedures that can be carried out with reasonable effort and are based only on reliable assumptions. Therefore, the aim of the investigation is the general solution of both the balance conditions for the forces and torques and the known boundary conditions.

1.1 The Calculation of Stresses

For a long time, there has been a well-known mathematical method that is suitable for calculating stresses in glaciers. However, it has never been applied in glaciology. With this method, the so-called weightless stress tensor fields can be represented.

These weightless stress tensor fields will be used to calculate stresses in the following. In order to describe the concept and the objective of this investigation, the relevant physical mechanisms are first presented and the weightless stress tensor fields are characterized.

1.2 The Physical Mechanisms

The following physical mechanisms govern stresses in glaciers [4, pp. 258–261]:

P. Halfar, *Stresses in glaciers*, https://doi.org/10.1007/978-3-662-66024-9_1

The Balance Conditions

The balance conditions for the forces and torques apply because every part of a glacier is free of acceleration. This is not exactly true, but it is almost always true to a very good approximation, since accelerations in glaciers are almost always negligible compared to the acceleration due to gravity. That's why you don't usually feel any acceleration forces when standing on a glacier and being carried along by its motion. This is because one's own body weight is much greater than the acceleration force acting on one's own body, because the acceleration due to gravity is much greater than the acceleration due to flow. On the other hand, in the case of tangibly jerky glacial movements, the accelerations cannot be neglected.

Thus, from a dynamic point of view, a glacier can be considered as an acceleration-free continuum, which is consequently in perfect static balance. This means that for any given part of a glacier, both the external forces and the external torques disappear altogether.

The Boundary Conditions

"Boundary conditions" are understood here as the reliably known boundary conditions, in contrast to the unknown boundary conditions mentioned below. These boundary conditions consist of the conditions of vanishing boundary stresses on free glacier surfaces and of the hydrostatic boundary conditions on boundary surfaces in standing waters. The atmospheric pressure is neglected.

The Unknown Boundary Conditions

Unknown or at least not reliably known boundary conditions occur at the subsurface and at the (fictitious) interfaces to the glacier area not considered.

Flow Conditions: Incompressibility and Flow Law

Incompressibility means that any material part of a glacier retains its volume during flow motion. Mathematically, this is expressed by the freedom from divergence of the velocity field of the flow motion.

The flow law describes a relation between the strain rates of the flow movement and the stresses.

1.3 The Weightless Stress Tensor Fields

The weightless stress tensor fields form the general solution of the homogenized balance conditions, where the specific ice weight is considered as a formal parameter and set to zero. Consequently, these weightless stress tensor fields can be interpreted as stress tensor fields in fictitious weightless glaciers.

The weightless stress tensor fields can be calculated by the method mentioned at the beginning, which has been known for a long time. With their help one can give the general solution of the balance conditions by adding a special solution of these balance conditions.

1.4 Concept and Aim of the Study

1.4.1 General

The general concept of investigation is to develop computational procedures that are based on reliable preconditions and can be performed with reasonable effort. The general solution of the balance and boundary conditions satisfies this concept. According to this concept, however, further conditions cannot be considered because the remaining boundary conditions are unknown and the flow conditions would result in an unjustifiably high computational effort.

Therefore, the general solution of the balance and boundary conditions is the aim of this investigation. The calculation of this general solution is essential content of the general part II of this book. Chapter 7 is its heart, because in this chapter the weightless stress tensor fields are constructed with boundary conditions, with the help of which one can immediately give the general solution. A summary description of the general solution and its possible applications is given in Chap. 9.

1.4.2 Special

In addition to the basic part I and the general part II, the book also contains a special part III. In this special part, not only properties and applications of the general solution are discussed, but also the question of how to select a realistic stress tensor field from the infinite number of stress tensor fields of the general solution is addressed. Since this selection problem cannot be solved in general due to its complexity, two aspects of this selection problem are treated exemplarily.

A practical aspect concerns the task of avoiding unnecessary computational effort in the calculation of a suitable solution, i.e. not to aim at a precision by too high effort, which cannot be achieved anyway due to uncertain preconditions. How to approach this task is demonstrated by the example of quasi-stagnant stress tensor fields in Sect. 10.3. These quasi-stagnant stress tensor fields can be calculated with relatively little effort and are candidates for realistic stress tensor fields in nearly stagnant glaciers.

A theoretical aspect of the selection problem concerns the question of the ideal solution, i.e. whether a unique solution can be defined by taking into account the complete balance, boundary and flow conditions and what this solution looks like. Such ideal solutions are calculated for horizontally infinitely extended, isotropic and homogeneous tabular iceberg models in Sect. 11.3.

2 Balance and Boundary Conditions

Abstract

The balance conditions for the forces and torques as well as the reliably known boundary conditions are formulated as conditions for the stress tensor field.

The glacier region under consideration is denoted by Ω its closed boundary by $\partial\Omega$.[1] The balance conditions state that each subregion ω of this glacier region is acceleration-free and consequently can be regarded as a rigid body which is in static balance [4, p. 258]. Thus, the forces and torques acting on this subregion ω cancel out altogether.[2] These forces and torques can be represented by integrals over the differential forces and torques acting on the differential ice masses inside the subregion ω and those acting from the outside on the oriented differential[3] surface elements of its boundary $\partial\omega$.

[1] It is assumed that the model glacier domain Ω has the simple topological structure of a sphere. Ω and the sphere can thus be mapped into each other by a continuous and invertible mapping, where the closed boundary $\partial\Omega$ and the sphere surface merge.

[2] In classical mechanics, freedom from acceleration of a system of point-like particles, which follow Newton's second and third laws of motion, is equivalent to the fact that for each subsystem the external forces and torques disappear altogether [1, pp. 5–7]. This characteristic property of acceleration-free systems is transferred to the continuum mechanics relevant for glaciers, since this is only a phenomenological variant of the classical mechanics of point-like particles, which follow Newton's second and third laws of motion.

[3] The orientation of a surface element is determined by the direction of its normal. This oriented normal is the unit vector pointing to the corresponding side and perpendicular to the area element. An oriented surface element consists of the surface element itself and its oriented normal. Here the normal points outwards.

P. Halfar, *Stresses in glaciers*, https://doi.org/10.1007/978-3-662-66024-9_2

The differential force acting due to the acceleration of gravity **g** on a differential mass of ice with differential volume dV and ice density ρ is equal to its differential weight (2.1). The differential force (2.2) acting on an oriented surface element with differential area dA is defined by the stress vector **Sn**, where **S** denote[4] the stress tensor and **n** the oriented surface normal. It is a surface force between the ice masses abutting the differential surface from both sides. This surface force is exerted from the ice mass on one side of the differential surface and acts through this differential surface on the ice mass on the other side. In the respective calculation (2.2) of these two surface forces, the surface normal **n** points to the ice mass from which the force is exerted. These two surface forces are opposite to each other because of the reversal of direction of the surface normal **n** and thus fulfil the principle "Actio equals Reactio". In Fig. 2.1, the stress vector **Sn** is drawn on the side of the surface from which the force, and thus the stress, originates. If the stress vector strikes the surface, its normal component is a compressive stress acting through this surface on the other side, otherwise it is a tensile stress. From the differential forces one obtains the differential torques by forming the vector products with the position vectors **r** (2.3) and (2.4).

The resultant force and the resultant torque acting on any subregion ω are obtained by integrating all differential forces (2.1) and (2.2) and torques (2.3) and (2.4) both inside the subregion ω and on its closed boundary ∂ω. These resulting quantities are supposed to vanish according to the balance conditions (2.8) and (2.9). In these balance conditions, the surface integrals over the closed boundary ∂ω are equal to the force or torque acting on this boundary. The volume integrals over the subregion ω are equal to the force or torque

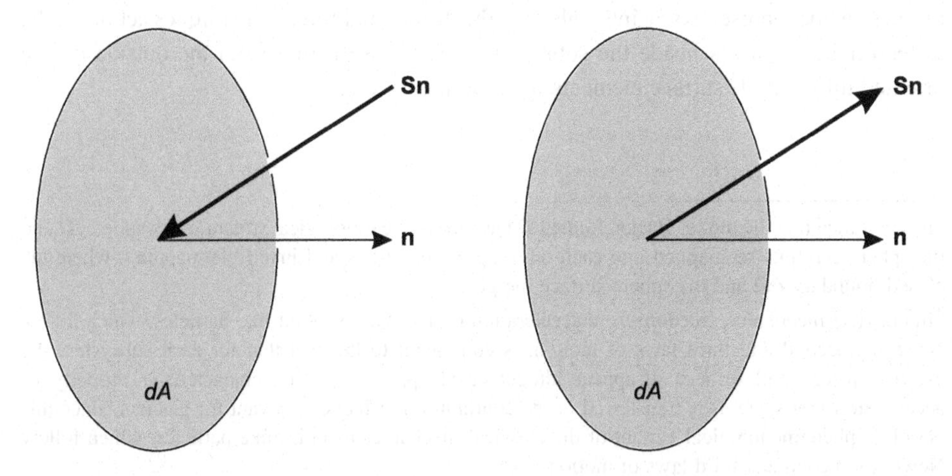

Fig. 2.1 Oriented differential surface elements dA with their oriented surface normals **n** and with stress vectors **Sn**

[4]For the justification that the stress vector can be written as **Sn**, see [5, pp. 134–135].

caused by the weight distribution in this subregion. These volume integrals can be expressed by the mass m_ω (2.5) or the moment $\mathbf{M}_\omega$ (2.6) of the mass distribution in the subregion ω and are equal to the weight of this mass $m_\omega$or equal to the torque of the mass fictitiously concentrated in its centre of gravity $\mathbf{c}_\omega$ (2.7).[5]

To convert the balance conditions from integral forms (2.8) and (2.9) to local forms, one converts[6] the integrals over the closed boundary $\partial\omega$ of the domain ω to volume integrals (2.10) and (2.11) over the domain ω using Gauss' theorem. Thus, the integral balance conditions can be expressed by pure volume integrals, which must vanish for any subrange ω of the glacier range Ω under consideration (2.12) and (2.13). Therefore, these integral balance conditions are equivalent to the local balance conditions, which consist of a differential equation (2.14) and the symmetry condition (2.15) for the stress tensor **S**.

For the boundary conditions (2.16) only reliably known data are considered, namely the boundary stresses **s** on that surface Σ of the glacier region Ω, which consists of its free surface and its boundary surface in stagnant waters. The boundary stresses on the free surface disappear and the boundary stresses in stagnant waters are given by the hydrostatic pressure $\widetilde{p}$ acting in opposite direction to the outward normal vector **n** of the surface Σ (2.17). The atmospheric pressure is neglected.[7]

The general solution **S** of the balance and boundary conditions (2.14)–(2.16) can be transformed into the general weightless solution **T** (2.20) of the simpler balance and boundary conditions (2.21)–(2.23) for weightless stress tensor fields by subtracting any special solution $\mathbf{S}_{\text{bal}}$ of the balance conditions (2.18) and (2.19).[8] The stress tensor fields are **T** called "weightless" because in the corresponding balance condition (2.21) for **T** the specific ice weight $\rho\mathbf{g}$ does not appear, unlike in the corresponding balance condition (2.14) for **S**.[9] The boundary stresses **t** of the weightless stress tensor fields **T** on the boundary surface Σ are the difference (2.24) of the known boundary stresses **s** and the boundary stresses of the tensor field $\mathbf{S}_{\text{bal}}$ and are thus also known. By transforming the local balance conditions (2.21) and (2.22) for weightless stress tensor fields **T** into their integral forms

[5] Instead of the moments $\mathbf{M}_\omega$ (2.6) in the areas ω one could also use the vectors $\mathbf{c}_\omega$ (2.7) from the origin of coordinates to the centers of mass. However, it is easier to work with these moments, because they add up in case of area extensions, whereas the centroid vectors do not. These moments $\mathbf{M}_\omega$ (2.6) and also the center of mass vectors $\mathbf{c}_\omega$ (2.7) depend on the position of the origin of coordinates.

[6] Let div**S** (13.8) denote the divergence of the tensor field **S** formed line by line, $\not{\mathbf{S}}$ (12.11) the vector field belonging to the skew-symmetric part of **S**. For the balance condition for the torques one writes $\mathbf{r} \times \mathbf{Sn}$ as $\not{\mathbf{r}}\mathbf{Sn}$ and reshapes the divergence of $\not{\mathbf{r}}\mathbf{S}$ according to (13.21).

[7] According to Archimedes' principle, the air pressure causes a weight reduction of the ice by buoyancy, which can be taken into account by interpreting the quantity ρ in the calculations not as ice density, but as the difference between ice density and air density. However, this change is in the per mille range and is therefore neglected.

[8] For the explanations in this chapter it is sufficient to know that the balance conditions (2.18), (2.19) are fulfilled by $\mathbf{S}_{\text{bal}}$, the solution itself need not be known. In Chap. 5 such a solution is constructed.

[9] More precisely, these **T** are stress tensor fields in fictitious weightless media. The chosen term "weightless stress tensor fields" is somewhat inaccurate, but not that cumbersome.

(2.25) and (2.26), it becomes clear that weightless stress tensor fields **T** do not produce resultant forces and torques on oriented closed surfaces, since the regions enclosed by these surfaces are weightless.

$$\rho \cdot \mathbf{g} \cdot dV \tag{2.1}$$

$$\mathbf{Sn} \cdot dA \tag{2.2}$$

$$\mathbf{r} \times \rho \cdot \mathbf{g} \cdot dV \tag{2.3}$$

$$\mathbf{r} \times \mathbf{Sn} \cdot dA \tag{2.4}$$

* * *

$$m_\omega \overset{\text{def.}}{=} \int_\omega \rho \cdot dV \tag{2.5}$$

$$\mathbf{M}_\omega \overset{\text{def.}}{=} \int_\omega \mathbf{r}\rho \cdot dV \tag{2.6}$$

$$\mathbf{c}_\omega \overset{\text{def.}}{=} \frac{\mathbf{M}_\omega}{m_\omega} \tag{2.7}$$

* * *

$$\oint_{\partial\omega} \mathbf{Sn} \cdot dA + \underbrace{\int_\omega \rho \cdot \mathbf{g} \cdot dV}_{m_\omega \mathbf{g}} = 0; \quad \omega \overset{\text{prec.}}{\subseteq} \Omega \tag{2.8}$$

$$\oint_{\partial\omega} \mathbf{r} \times \mathbf{Sn} \cdot dA + \underbrace{\int_\omega \rho \cdot \mathbf{r} \times \mathbf{g} \cdot dV}_{\mathbf{M}_\omega \times \mathbf{g} = \mathbf{c}_\omega \times m_\omega \mathbf{g}} = 0; \quad \omega \overset{\text{prec.}}{\subseteq} \Omega \tag{2.9}$$

$$\oint_{\partial\omega} \mathbf{Sn} \cdot dA \overset{\text{id.}}{=} \int_\omega \operatorname{div} \mathbf{S} \cdot dV \tag{2.10}$$

$$\oint_{\partial\omega} \mathbf{r} \times \mathbf{Sn} \cdot dA \overset{\text{id.}}{=} \int_\omega [\mathbf{r} \times \operatorname{div} \mathbf{S} + 2 \cdot \$] \cdot dV \tag{2.11}$$

$$\int_{\omega} (\text{div}\,\mathbf{S} + \rho\mathbf{g}) \cdot \text{d}V = \mathbf{0}; \quad \omega \overset{\text{prec.}}{\subseteq} \Omega \tag{2.12}$$

$$\int_{\omega} [\boldsymbol{r} \times (\text{div}\,\mathbf{S} + \rho\mathbf{g}) + 2 \cdot \mathbf{\$}] \cdot \text{d}V = \mathbf{0}; \quad \omega \overset{\text{prec.}}{\subseteq} \Omega \tag{2.13}$$

$$\text{div}\,\mathbf{S} + \rho\mathbf{g} = \mathbf{0} \tag{2.14}$$

$$\mathbf{S} = \mathbf{S}^{\text{T}} \tag{2.15}$$

$$\mathbf{S}|_{\Sigma} \cdot \mathbf{n} = \mathbf{s} \tag{2.16}$$

$$\mathbf{s} \overset{\text{def}}{=} \begin{cases} \mathbf{0} & \text{on free surfaces} \\ -\tilde{p} \cdot \mathbf{n} & \text{in standing waters} \end{cases} \tag{2.17}$$

* * *

$$\text{div}\,\mathbf{S}_{\text{bal}} + \rho\mathbf{g} = \mathbf{0} \tag{2.18}$$

$$\mathbf{S}_{\text{bal}} = \mathbf{S}_{\text{bal}}^{\text{T}} \tag{2.19}$$

$$\mathbf{T} \overset{\text{def.}}{=} \mathbf{S} - \mathbf{S}_{\text{bal}} \tag{2.20}$$

$$\text{div}\,\mathbf{T} = \mathbf{0} \tag{2.21}$$

$$\mathbf{T} = \mathbf{T}^{\text{T}} \tag{2.22}$$

$$\mathbf{T}|_{\Sigma} \cdot \mathbf{n} = \mathbf{t} \tag{2.23}$$

$$\mathbf{t} \overset{\text{def}}{=} \mathbf{s} - \mathbf{S}_{\text{bal}}|_{\Sigma} \cdot \mathbf{n} \tag{2.24}$$

$$\oint_{\partial\omega} \mathbf{Tn} \cdot \text{d}A = \mathbf{0}; \quad \omega \overset{\text{prec.}}{\subseteq} \Omega \tag{2.25}$$

$$\oint_{\partial\omega} \mathbf{r} \times \mathbf{Tn} \cdot \text{d}A = \mathbf{0}; \quad \omega \overset{\text{prec.}}{\subseteq} \Omega \tag{2.26}$$

Integral Operators

3

Abstract

If conventional integrations were used, the formulas would become confusing and the calculations cumbersome. With the integral operators introduced here, formulas can be made clear and calculations can be performed easily.

3.1 Example, General Properties and Minimal Models

Compared to conventional integrals, integral operators make formulas clearer and calculations easier to perform.

These advantages of integral operators shall be demonstrated by a typical example. A function *f* is integrated over a path leading from a point in the glacier in *z*-direction to its free surface. The result does not change if this function *f* disappears above the free surface (3.1) and one continues the integration to infinity (3.2). If one reverses the sign, one obtains an arithmetic operation (3.3) which is inverse to the differentiation and is therefore symbolized by ∂_z^{-1}.[1] This integral operator ∂_z^{-1} is interchangeable with all differential operators (3.4) and its multiplication by the differential operator ∂_z gives 1 (3.5). The repeated application of this integral operator ∂_z^{-1} can be defined as a negative power of the differential operator ∂_z (3.6), so that all integer powers of the differential operator ∂_z are defined, where the zeroth power is said to be equal to 1, as usual (3.7). For the multiplication of these integer powers, the familiar rules of calculation apply (3.8). These rules of interchange (3.4) and power (3.8) for the integral operators make the calculation quite easy,

[1] The differential operator ∂_z is invertible because only functions that vanish above the free surface are allowed, which fixes the integration constant.

P. Halfar, *Stresses in glaciers*, https://doi.org/10.1007/978-3-662-66024-9_3

in contrast to the cumbersome form (3.9) of the rules of interchange and the even more cumbersome form of the power rules in conventional notation.

To demonstrate the equivalence between the interchange rules (3.4) for the integral operator ∂_z^{-1} and the conventional interchange rules (3.9), write the function f as the product (3.10) of a function f_c and a step function. The function f_c is continuous and differentiable everywhere and coincides with the function f in the glacier region. The step function vanishes above the free surface and otherwise has the value 1 and ensures that the function f vanishes above the free surface. This step function can be represented using the Heaviside function θ (3.11), which vanishes for negative arguments and has value 1 otherwise. Thus it can be shown that transformations (3.13) using the permutation rules (3.4) for the integral operator ∂_z^{-1} lead to the same result as transformations (3.9) using the conventional permutation rules. The derivative of the Heaviside function occurs, which is the delta function δ (3.12).

Several integral operators may also occur, for example ∂_z^{-1} and ∂_y^{-1}, where these are applied only to functions which admit the convergence of the corresponding integrals, and these various integral operators are interchangeable with each other as well as with all differential operators.

From this example we can already recognize some of the following characteristic properties, which are common to all integral operators appearing here:

1. Integral operators can be used to make formulas clear and to perform calculations easily. The class of so-called "admissible functions and distributions" must be specified, on which the integral operators may act through integrations and from which they generate other functions. This class of admissible functions and distributions also includes the delta function and its derivatives, which are created by derivatives of functions with discontinuities.
2. The admissible functions and distributions are defined in a range larger than the glacier range under consideration, which contains the so-called "range of vanishing function values" on which all admissible functions and distributions must vanish. All integrations defined by integral operators extend over ranges that lie almost entirely[2] in the range of vanishing function values, which guarantees the convergence of all integrals.
3. The class of admissible functions and distributions is closed under linear combinations and under the applications of the occurring integral operators as well as the applications of all differential operators.[3] Therefore one can form arbitrary linear combinations and apply the integral and all differential operators arbitrarily often, so that many design possibilities arise.
4. The following rules simplify the calculations: All integral operators that occur are the inverses of differential operators, and the product of an integral operator and its inverse

[2] "Almost entirely" means "entirely except for finite portions".

[3] This means that these operations generate admissible functions and distributions.

differential operator equals one. For the occurring integral operators and all differential operators, the commutative law applies.[4]

5. The restriction to admissible functions and distributions is not intended to exclude any functions that might be necessary to describe the stresses. We want to assume that this includes all smooth (arbitrarily often differentiable) functions in the considered glacier domain.[5] The domain of definition of admissible functions and distributions will generally be larger than the glacier domain under consideration,[6] and functions that can be differentiated arbitrarily many times on the glacier domain under consideration will be continued into the remaining domain of definition by the value zero.

For each model with these properties there is the minimal model, which is defined by the minimal class of admissible functions and distributions. This minimal class consists of the functions which are smooth in the considered glacier area and vanish outside of it, as well as of all functions and distributions which are generated from it by arbitrary polynomials of the occurring integral operators and all differential operators.[7]

—

$$f(x, y, z) = 0; \quad z > z_0(x, y) \tag{3.1}$$

$$\int_z^{z_0(x,\, y)} \mathrm{d}z' \cdot f(x, y, z') = \int_z^{\infty} \mathrm{d}z' \cdot f(x, y, z') \tag{3.2}$$

* * *

$$\partial_z^{-1} f \overset{\text{def.}}{=} -\int_z^{\infty} \mathrm{d}z' \cdot f(x, y, z') \tag{3.3}$$

$$\partial_i \partial_z^{-1} = \partial_z^{-1} \partial_i; \quad i = x{,}y{,}z \tag{3.4}$$

$$\partial_z \partial_z^{-1} = \partial_z^{-1} \partial_z = 1 \tag{3.5}$$

[4] All operators mentioned are interchangeable.

[5] In practice, one can get by with even smaller function classes, for example, by not assuming all smooth functions, but only all polynomials.

[6] See the example in Sect. 3.1.

[7] This class of admissible functions and distributions is called minimal, because every other class according to the properties mentioned above in points 3 and 5 contains this minimal class.

$$\partial_z^{-m} \overset{\text{def.}}{=} \left(\partial_z^{-1}\right)^m; \quad m = 1,2,\ldots \tag{3.6}$$

$$\partial_z^0 \overset{\text{def.}}{=} 1 \tag{3.7}$$

$$\partial_z^m \partial_z^n = \partial_z^{m+n}; \quad m,n = \ldots, -1,0,1,\ldots \tag{3.8}$$

***

$$\partial_i \int_z^{z_0} \mathrm{d}z' \cdot f = \int_z^{z_0} \mathrm{d}z' \cdot \partial_i f + [f]_{z=z_0} \cdot \partial_i (z_0 - z); \quad i = x,y,z \tag{3.9}$$

$$f(x, y, z) = f_c(x, y, z) \cdot \theta(z_0 - z) \tag{3.10}$$

$$\theta(z) \overset{\text{def.}}{=} \begin{cases} 1; & 0 \le z \\ 0; & z < 0 \end{cases} \tag{3.11}$$

$$\partial_z \theta(z) = \delta(z) \tag{3.12}$$

$$\begin{aligned} \partial_i \int_z^{z_0} \mathrm{d}z' \cdot f &= -\partial_i \partial_z^{-1} [f_c \cdot \theta(z_0 - z)] \\ &= -\partial_z^{-1} \partial_i [f_c \cdot \theta(z_0 - z)] \\ &= -\partial_z^{-1} [\theta(z_0 - z) \cdot \partial_i f_c + f_c \cdot \delta(z_0 - z) \cdot \partial_i (z_0 - z)] \\ &= \int_z^{z_0} \mathrm{d}z' \cdot \partial_i f_c + [f_c]_{z=z_0} \cdot \partial_i (z_0 - z); \quad i = x,y,z \end{aligned} \tag{3.13}$$

3.2 Integral Operators

Only the integral operators denoted $(\mathbf{a}\nabla)^{-1}$ and $\Box_z^{-1}$ occur, which are the inverses of the differential operators $\mathbf{a}\nabla$ (3.14) and $\Box_z$ (3.16).[8] These integral operators are each characterized by their convex integration cone[9] and a weight function defined on that cone, and generate a new function from a function by assigning to each point the integral

[8] In addition, there are the cases which result from renaming the location coordinates.

[9] The term "convex cone" denotes a geometric structure with a defined shape and orientation. This term is used both concretely to designate such a structure at a certain position in space, which is defined by the position of the cone's apex, and abstractly without taking its position into account.

over the integration cone starting from that point,[10] with the integrand containing the original function multiplied by the weight function. The integral operator $(\mathbf{a}\nabla)^{-1}$ (3.17) has a one-dimensional integration cone generated by the vector **a**.[11] Its weight function is constant and given by the negative inverse length of the vector **a**. The integral operator $\Box_z^{-1}$ (3.19) has a three-dimensional, rotationally symmetric integration cone, denoted by $K_z^{\odot}$, whose rotation semi-axis points in the positive z-direction and whose opening angle is equal to a right angle. The weight function G (3.20) of the integral operator $\Box_z^{-1}$ has an integrable singularity on the cone surface.[12]

If there are some integral operators of this type, there is always a model and therefore a minimal model with the required properties, if these integral operators satisfy the following condition[13]:

- There is an oriented plane such that this plane and each integration cone of the model are each transverse and synchronous to each other.

By definition, a cone and an oriented plane are transverse to each other if all straight lines parallel to the cone rays intersect the plane exactly once. They are synchronous to each other if the passage of the cone rays through the plane is in the direction of the oriented normal of the plane.

Thus, all integration cones of the model are oriented in such a way that they are asymptotically contained in the half-space[14] bounded by the named plane and into which the oriented normal of this plane points.

In order to show that under this condition, for example, the minimal model[15] exists, one shifts the plane mentioned in the condition and the half-space bounded by it, which contains all integration cones asymptotically, parallel so far in the direction of the cone rays that this half-space lies completely outside the glacier area considered. Thus, in this half-space the functions which can be differentiated arbitrarily often in the glacier region under consideration and which otherwise disappear, and thus all functions which are

[10]The apex of this cone is at this point.

[11]In the following, the integral operator $(-\mathbf{a}\nabla)^{-1}$ is also defined by $-(\mathbf{a}\nabla)^{-1}$ Thus an integral operator $(\mathbf{b}\nabla)^{-1}$ is defined not until the cone vector is defined, which is either **b** or $-\mathbf{b}$.

[12]This weight function $G(\mathbf{r}' - \mathbf{r})$ (3.20) on the integration cone with apex in the point **r** is a function of the cone vectors $\mathbf{r}' - \mathbf{r}$, which lead from the apex in the point **r** to the other points $\mathbf{r}'$ of the cone. The function $G(\mathbf{r}' - \mathbf{r})$ is also defined for points $\mathbf{r}'$outside the cone, where it is vanishing. Therefore, the integration (3.19) formally extends over the whole space.

[13]This condition, which is sufficient for the existence of models with properties 1 to 5 in Sect. 3.1, is probably also necessary, but this will not be investigated further. It applies to every model in this paper (but not simultaneously to all models).

[14]This means that in each case the whole integration cone lies in the half-space if its apex lies in it and that the cone extends into the half-space if its apex does not lie in the half-space.

[15]See Sect. 3.1.

generated from these functions by the integral and all differential operators, i.e. all admissible functions and distributions of the minimal model, disappear. All integrals defined by the integral operators over the integration cones converge, since all integration cones are asymptotically contained in this half-space of vanishing function values. It follows from this construction of the minimal model that it has the required properties,[16] except for the interchangeability of all operators. That it also has this property, that is, that the integral and all differential operators are interchangeable, follows from the interchangeability of all differential operators and from the fact that the integral operators are the unique inverses (3.21) of corresponding differential operators. Their unique invertibility is enforced by the fact that all admissible functions and distributions must vanish in a half-space in which all integration cones are asymptotically contained.[17]

If all integration cones are asymptotically contained in a half-space, then this is also true for the so-called model cone, which is generated by all integration cones occurring in the respective model.[18] Thus, the integral operators $(\mathbf{k}\nabla)^{-1}$ are defined for all cone vectors $\mathbf{k}$ of the model cone. If the integral operators $(\mathbf{k}\nabla)^{-1}$ or $\square_z^{-1}$ are defined, then the integral operators $(-\mathbf{k}\nabla)^{-1}$ or $(-\square)_z^{-1}$ are also defined (3.22) by changing the sign of the weight function in each case, while the integration cone remains the same.[19] For the basis vectors, the agreement is that all basis vectors that are parallel to a conic ray of the model cone are oriented so that they are also conic vectors. Thus negative basis vectors can never be cone vectors of the model cone. Powers of integral operators are negative powers (3.24) of the corresponding differential operators, where the 0th powers are equal to 1 (3.23). All operators are interchangeable (3.25) and known rules of calculation apply to powers of operators (3.26) and (3.27).[20]

From these explanations it follows that, when constructing a model, one must pay particular attention to the fact that there is a half-space in which all integration cones of the model, and thus also the model cone generated by these integration cones, are asymptotically contained. This provides the minimal model,[21] which is usually sufficient for modeling real situations. This model contains the minimal class of admissible functions and distributions, which consists of all functions that can be differentiated arbitrarily often

---

[16] These are the characteristics listed in Sect. 3.1 under points 1 to 5.

[17] The proof of unique invertibility (3.21) is as easy to give for $\mathbf{a}\nabla$ as in footnote 1 in Sect. 3.1 for ∂_z. For $\square_z$ this proof follows from the theory of hyperbolic differential equations.

[18] All generating cones are convex. A generated convex cone is the smallest convex cone that contains all generating convex cones. Its cone vectors consist of all linear combinations of cone vectors of the generating cones with non-negative coefficients. The cone vectors of a convex cone are the vectors leading from the cone apex to the cone points.

[19] See footnote 11.

[20] The calculation rules (3.25)–(3.27) apply to all differential operators and all their integer powers, but with the restriction that a differential operator with negative power may only occur if its (-1)-th power belongs to the model as an integral operator.

[21] See Sect. 3.1.

in the considered glacier domain and vanish outside of it, and furthermore of the functions and distributions that can be generated from it by arbitrary polynomials of the occurring integral operators and all differential operators. This includes generalized functions such as delta functions and their derivatives, which arise by differentiating ordinary functions with discontinuities. Arbitrary polynomials of integral and all differential operators are defined on this minimal class of admissible functions and distributions and again generate functions from this class. This results in many possibilities of design, which are easy to calculate with, because all operators are interchangeable (3.25) and because the rules of calculation (3.26) and (3.27) are valid for powers of operators.

—

$$\mathbf{a}\nabla = a_x\partial_x + a_y\partial_y + a_z\partial_z \tag{3.14}$$

$$\begin{aligned}[(\mathbf{a}\nabla)f](\mathbf{r}) &= (a_x\partial_x + a_y\partial_y + a_z\partial_z)f(\mathbf{r}) \\ &= \mid \mathbf{a} \mid \cdot \left[\partial_s f\left(\mathbf{r} + s\frac{\mathbf{a}}{\mid \mathbf{a} \mid}\right)\right]_{s=0} \\ &= [\partial_\alpha f(\mathbf{r} + \alpha \cdot \mathbf{a})]_{\alpha=0}\end{aligned} \tag{3.15}$$

$$\Box_z \stackrel{\text{def}}{=} \partial_z^2 - \partial_x^2 - \partial_y^2 \Box_z \tag{3.16}$$

* * *

$$\begin{aligned}\left[(\mathbf{a}\nabla)^{-1}f\right](\mathbf{r}) &\stackrel{\text{def.}}{=} -\int_0^\infty \mathrm{d}\alpha \cdot f(\mathbf{r} + \alpha\mathbf{a}) \\ &= -\frac{1}{\mid \mathbf{a} \mid}\int_0^\infty \mathrm{d}s \cdot f\left(\mathbf{r} + s\frac{\mathbf{a}}{\mid \mathbf{a} \mid}\right)\end{aligned} \tag{3.17}$$

$$s = \mid \alpha \cdot \mathbf{a} \mid \tag{3.18}$$

$$\left[\Box_z^{-1}f\right](\mathbf{r}) \stackrel{\text{def}}{=} \int \mathrm{d}x'\mathrm{d}y'\mathrm{d}z' \cdot G(\mathbf{r}' - \mathbf{r}) \cdot f(\mathbf{r}') \tag{3.19}$$

$$G(\mathbf{r}' - \mathbf{r}) \stackrel{\text{def.}}{=} \frac{1}{2\pi} \cdot \frac{\theta\left[(z' - z) - \sqrt{(x' - x)^2 + (y' - y)^2}\right]}{\sqrt{(z' - z)^2 - (x' - x)^2 - (y' - y)^2}}. \tag{3.20}$$

* * *

$$(\mathbf{a}\nabla)\cdot(\mathbf{a}\nabla)^{-1} = (\mathbf{a}\nabla)^{-1}\cdot(\mathbf{a}\nabla) = \Box_z\cdot\Box_z^{-1} = \Box_z^{-1}\cdot\Box_z = 1 \tag{3.21}$$

* * *

$$(-\mathbf{a}\nabla)^{-1} \overset{\text{def}}{=} -(\mathbf{a}\nabla)^{-1};\quad (-\Box_z)^{-1} \overset{\text{def}}{=} -\Box_z^{-1} \tag{3.22}$$

$$(\mathbf{a}\nabla)^0 \overset{\text{def}}{=} 1;\quad \Box_z^0 \overset{\text{def}}{=} 1 \tag{3.23}$$

$$(\mathbf{a}\nabla)^{-\mathrm{m}} \overset{\text{def}}{=} [(\mathbf{a}\nabla)^{-1}]^{\mathrm{m}};\quad \Box_z^{-\mathrm{m}} \overset{\text{def}}{=} (\Box_z^{-1})^{\mathrm{m}};\quad m = 1,2,\ldots \tag{3.24}$$

* * *

$$(\mathbf{a}\nabla)^m(\mathbf{b}\nabla)^n = (\mathbf{b}\nabla)^n(\mathbf{a}\nabla)^m;\quad (\mathbf{a}\nabla)^m\cdot\Box_z^n = \Box_z^n\cdot(\mathbf{a}\nabla)^m;\quad m,n = \cdots -1,0,1,\ldots \tag{3.25}$$

$$(\mathbf{a}\nabla)^m(\mathbf{a}\nabla)^n = (\mathbf{a}\nabla)^{n+m};\quad \Box_z^n\cdot\Box_z^m = \Box_z^{n+m};\quad m,n = \cdots -1,0,1,\ldots \tag{3.26}$$

$$(\lambda\cdot\mathbf{a}\nabla)^m = \lambda^m\cdot(\mathbf{a}\nabla)^m;\quad (\lambda\cdot\Box_z)^m = \lambda^m\cdot\Box_z^m;\quad \lambda \neq 0 \tag{3.27}$$

3.3 Dependency Cones and Products of Integral Operators

If one applies an integral operator to a function, then the result at any point depends only on the function values in the integration cone starting from that point. Therefore, the convex integration cone of an integral operator is also called its "dependence cone". A product of integral operators also has a convex dependence cone and this is generated by the convex dependence cones of the factors.[22] This dependence cone of a product is transverse and synchronous to the oriented plane to which the dependence cones of the factors are also transverse and synchronous.[23]

Unlike the integral operators themselves, a product of integral operators cannot, as a rule, be represented as an integral over its dependency cone. An exception is made for products whose factors are of type $(\mathbf{a}\nabla)^{-1}$ where the vectors $\mathbf{a},\ldots$ of the factors are linearly

[22] See footnote 18.

[23] See the condition in Sect. 3.2.

independent and lie in the model cone. The integral operator $(\mathbf{a}\nabla)^{-1}$ (3.17) causes a path integration over the one-dimensional cone generated by the vector **a**, divided by the negative length of the vector **a**. The product $(\mathbf{a}\nabla)^{-1}(\mathbf{b}\nabla)^{-1}$ (3.28) causes an area integration over the two-dimensional cone generated by its boundary vectors **b** and **a** divided by the area of the parallelogram generated by **a** and **b**. The product $(\mathbf{a}\nabla)^{-1}(\mathbf{b}\nabla)^{-1}(\mathbf{c}\nabla)^{-1}$ (3.30) causes a volume integration over the three-dimensional cone generated by its edge vectors **a**, **b** and **c** divided by the negative volume of the spade generated by **a**, **b** and **c**.

—

$$\left[(\mathbf{a}\nabla)^{-1}(\mathbf{b}\nabla)^{-1}f\right](\mathbf{r}) = \int_0^\infty\int_0^\infty \mathrm{d}\alpha\cdot\mathrm{d}\beta\cdot f(\mathbf{r}+\alpha\mathbf{a}+\beta\mathbf{b})$$

$$= \frac{1}{|\mathbf{a}\times\mathbf{b}|}\int \mathrm{d}A'\cdot f(\mathbf{r}+\alpha\mathbf{a}+\beta\mathbf{b}) \qquad (3.28)$$

$$\mathrm{d}A' = |\,\mathrm{d}\alpha\cdot\mathbf{a}\times\mathrm{d}\beta\cdot\mathbf{b}\,| \qquad (3.29)$$

$$\left[(\mathbf{a}\nabla)^{-1}(\mathbf{b}\nabla)^{-1}(\mathbf{c}\nabla)^{-1}f\right](\mathbf{r}) = -\int_0^\infty\int_0^\infty\int_0^\infty \mathrm{d}\alpha\cdot\mathrm{d}\beta\cdot\mathrm{d}\gamma\cdot f(\mathbf{r}+\alpha\mathbf{a}+\beta\mathbf{b}+\gamma\mathbf{c})$$

$$= -\frac{1}{|(\mathbf{a}\times\mathbf{b})\cdot\mathbf{c}|}\int \mathrm{d}V'\cdot f(\mathbf{r}+\alpha\mathbf{a}+\beta\mathbf{b}+\gamma\mathbf{c}) \qquad (3.30)$$

$$\mathrm{d}V' = |\,(\mathrm{d}\alpha\cdot\mathbf{a}\times\mathrm{d}\beta\cdot\mathbf{b})\cdot\mathrm{d}\gamma\cdot\mathbf{c}\,| \qquad (3.31)$$

3.4 Solutions of Boundary Value Problems of Partial Differential Equations

Each of the boundary value problems of linear, partial differential equations of second order that appear here consists in finding as a solution a function that satisfies both the differential equation and the boundary conditions. These boundary conditions consist of vanishing boundary values on a boundary surface Σ for the function itself and for its normal derivative. It is necessary to discuss such boundary value problems because they play a role in some representations of the general solution of the balance and boundary conditions

(2.14)–(2.16). The solutions of these boundary value problems can be represented by integral and differential operators. This shall be outlined in the following.[24]

3.4.1 Boundary Surfaces with Boundary Conditions, Domains of Definition and Minimal Models

Only homogeneous boundary conditions occur.[25] These are valid on a boundary surface Σ, which is here the free surface of the considered glacier area. A boundary value problem has a unique solution if both the shape of the free surface Σ and the glacier area under consideration Ω satisfy conditions characterized by the model cone. This model cone is convex and is generated by the convex integration cones of all integral operators[26] needed to solve the boundary value problem. The stated conditions are:

1. The model cone and the oriented free surface Σ are transverse and synchronous to each other. A cone and an oriented surface Σ are by definition transverse to each other if all straight lines parallel to the cone rays intersect the surface at most once. They are synchronous to each other if the passage of the cone rays through the oriented surface Σ is towards the side to which the oriented normal of the surface points. The oriented normal of the free surface Σ should be directed outwards.
2. The considered glacier region Ω is compatible with the model cone and the oriented boundary surface Σ. This means that any cone ray of the model cone emanating from this region Ω is uninterrupted in the region until it meets the boundary surface Σ.

If these conditions are fulfilled, the minimal model is available for the solution of the boundary value problem with the interchangeability rules (3.25) and power rules (3.26) and (3.27) for the integral operators and for all differential operators.[27] The domain of definition Ω_{def} of this minimal model is obtained by adding all model cones emanating[28] from the glacier domain Ω. Thus this domain of definition Ω_{def} of the minimal model consists of the glacier domain Ω and the so-called external domain Ω_{ext} beyond its free surface Σ, which is

[24] Such boundary value problems will appear in Sect. 8.2. There and in Chap. 17 the detailed formulations and solutions are given, while here a sketch is sufficient.

[25] The functions and their first derivatives vanish on the boundary surface Σ.

[26] See Sect. 3.2.

[27] The condition in Sect. 3.2 that there is an oriented plane such that this plane and each integration cone of the model are transverse and synchronous to each other is fulfilled. This is because each tangent plane of the boundary surface Σ represents such a plane. Therefore the minimal model is available, which is characterized in Sect. 3.1.

[28] These are all model cones with tip in Ω.

covered by all model cones starting from this free surface Σ.[29] In the external domain Ω_{ext} all admissible functions and distributions of the minimal model disappear. The class of these admissible functions and distributions contains all functions which are arbitrarily differentiable in the glacier domain Ω and vanish in the external domain Ω_{ext} beyond Σ. It also contains the functions and distributions which can be generated from the previously mentioned functions by arbitrary polynomials of the occurring integral operators and all differential operators.

This minimal model, which is available under the conditions given in (1) and (2) above, provides the framework in which the unique solution of the boundary value problem can be represented by integral and differential operators.

3.4.2 Boundary Value Problems in Minimal Models

Homogeneous boundary value problems arise for systems of three differential equations each with three sought functions. To represent such a boundary value problem clearly, one writes it in matrix form by defining three sought functions as three components of a column matrix function **f** and using a quadratic matrix operator $\mathcal{L}$ whose nine matrix elements each consist of second order differential operators. A boundary value problem in matrix form consists of determining, for a column matrix function **q** given on a glacier domain Ω that can be differentiated any number of times, a column matrix function **f** that satisfies a second-order matrix differential equation with the matrix operator $\mathcal{L}$ (3.32) and that satisfies homogeneous boundary conditions at the free surface Σ (3.33).

If one continues column matrix functions **f** and **q** defined initially only on the glacier domain Ω by the value zero into the external area Ω_{ext} beyond the free surface Σ, then the differential equation (3.32) is valid also on the whole definition area Ω_{def}, because the searched function **f** together with its first derivatives is continuous on the whole domain of definition Ω_{def} because of the homogeneous boundary conditions (3.33).[30] Thus the boundary value problem is in an equivalent form and consists only of a matrix differential equation (3.34), but on the whole domain of definition Ω_{def}. This boundary value problem is to be solved within the minimal model, where both the given function **q** and the function **f** we are looking for must vanish in the external domain Ω_{ext} beyond the free surface Σ.

---

[29] In the minimal model, one could take the entire space as the domain of definition. However, the domain of definition Ω_{def} is defined more narrowly here, since only this narrower definition range is important due to the problem.

[30] Because of this continuity of **f** and its first derivatives on Σ, the second derivatives may be discontinuous but without delta function-like components on Σ.

$$\mathcal{L}\mathbf{f} = \mathbf{q}; \quad \mathbf{r} \in \Omega \tag{3.32}$$

$$\mathbf{f}|_\Sigma = 0; \quad \partial_n \mathbf{f}|_\Sigma = 0 \tag{3.33}$$

$$\mathcal{L}\mathbf{f} = \mathbf{q}; \quad \mathbf{r} \in \Omega_{def} \stackrel{def}{=} \Omega \cup \Omega_{ext} \tag{3.34}$$

3.4.3 Solutions of Boundary Value Problems by Integral and Differential Operators

The solution of the matrix-differential equation (3.34) on the domain of definition Ω_{def} vanishes beyond the free surface Σ. In this case, boundary conditions are no longer imposed, but the solution of the matrix-differential equation (3.34) calculated in the domain of definition Ω_{def} automatically satisfies the homogeneous boundary conditions (3.33) on the free surface Σ.

To solve[31] the matrix differential equation (3.34) one constructs the operator inverse $\mathcal{L}^{-1}$ of the matrix operator $\mathcal{L}$ from the adjoint matrix operator $\mathcal{L}_{adj}$, whose elements consist of fourth order differential operators and from the determinant $\det(\mathcal{L})$, which is a sixth order differential operator. This determinant consists of factors of invertible differential operators of type $\mathbf{a}\nabla$ or $\Box_z$ and thus can also be inverted. Its inverse $[\det(\mathcal{L})]^{-1}$ is a product of integral operators of type $(\mathbf{a}\nabla)^{-1}$ or $\Box_z^{-1}$. Thus the operator $\mathcal{L}^{-1}$ (3.35) inverse to the matrix operator $\mathcal{L}$ can be given and the solution $\mathbf{f}$ (3.36) of the matrix differential equation (3.34) can be calculated.[32]

A solution $\mathbf{f}$ constructed in this way also Σ satisfies the homogeneous boundary conditions (3.33) at the free surface Σ, since this solution consists of products of integral and differential operators acting on admissible functions, producing zeros of at least second order at the free surface Σ. This can be explained as follows:

An integral operator of the type $(\mathbf{a}\nabla)^{-1}$ produces a first-order zero at the free surface Σ or raises the order of a zero by one, and the integral operator $\Box_z^{-1}$ counts twice because it

[31] This solution method is used in Sect. 8.2 and in Sects. 17.1, 17.2, 17.3, 17.4, 17.5, 17.6, 17.7, and 17.8. This is not a general treatise on boundary value problems of matrix differential equations, but only the matrix differential equations are studied and solved which are relevant for this treatise.

[32] $\mathcal{L}_{adj}$, and $\det(\mathcal{L})$ are uniquely defined, since all differential operators commutate with each other. Only those operators $\mathcal{L}$ will occur whose determinant $\det(\mathcal{L})$ is a product of invertible differential operators. All expressions are well defined, because all differential operators and all occurring inverted differential operators – that are the occurring integral operators – commutate with each other.

raises the order of a zero by two or produces a second-order zero.[33] In contrast, any first-order differential operator decreases the order of a zero, but by at most one.[34] Since in the operator products of the solution the integral operators occur in corresponding excess, these operator products each produce a zero of at least second order on the free surface Σ.

—

$$\mathcal{L}^{-1} = [\det(\mathcal{L})]^{-1} . \mathcal{L}_{\text{adj}} \tag{3.35}$$

$$\mathbf{f} = \mathcal{L}^{-1} \cdot \mathbf{q}; \quad \mathbf{r} \in \Omega_{\text{def}} \tag{3.36}$$

3.5 Integrations of Distributions with Integral Operators

The integral operators were previously defined only on functions. The calculation rules[35] for integral and differential operators are valid for arbitrary products of integral and differential operators, whereby the order of the operators is not important.[36] Because the differential operators can produce distributions from the admissible functions, which are no ordinary functions,[37] the mentioned rules of calculation are only valid without restrictions, if the integral operators are defined on all admissible distributions. This will be done in this chapter.

3.5.1 Domain of Definition

The requirements on the spatial domain of definition Ω_{def} of the distributions are influenced by the convex integration cones of the integral operators occurring in each case. The totality of these integral operators is characterized by the fact that there is an oriented plane

[33] This is especially true when approaching the free surface Σ from the glacier area, because beyond the free surface all permissible functions disappear anyway.

[34] The order of the zero is therefore at most decreased by one, because the differentiation can take place not only transversely, but also parallel to the boundary surface Σ, which as a rule does not result in a decrease of the zero order. On the other hand, the integral operators always lead to an increase of the zero order, because the integration cones always lie transverse to the boundary surface Σ.

[35] S. (3.21)–(3.27).

[36] A product of operators means that these operators are to be used one after the other. Here the order is not important, because all operators commutate with each other.

[37] In the following, the ordinary functions admissible here – that is, the functions smooth in the considered glacier domain Ω and vanishing in Ω_{ext} – are also called distributions, so that the class of all admissible distributions is an extension of the class of admissible ordinary functions.

to which all convex integration cones are transverse and synchronous. These convex integration cones generate the model cone, which is also convex and is the smallest convex cone in which all convex integration cones are contained.[38]

The spatial domain of definition Ω_{def}[39] of the distributions shall be such that arbitrary products of the integral operators are defined. Therefore, the domain of definition Ω_{def} shall contain all model cones starting from it, which furthermore all run into a so-called external subspace Ω_{ext} of the domain of definition.[40] In this outer subspace Ω_{ext} all admissible distributions vanish. Ω_{ext} also contains all model cones starting from it.

Under these conditions, all integrations can be performed, all integrations converge, and arbitrary products of integral operators are defined:

- All integrations can be performed, since all integration cones starting from the definition domain lie in the model cone and thus in the definition domain.
- All integrations converge as the integration cones run into the external subdomain Ω_{ext} of vanishing distributions.
- Arbitrary products of integral operators are defined, because every integral operator generates such distributions again from distributions which vanish in the external subdomain Ω_{ext}.

This is initially only true for distributions that are functions, but can be generalized to all distributions after defining the application of integral operators to distributions in the following.

Although the external subrange Ω_{ext} does not contain informations because of the vanishing of all distributions, it is needed for the definition of the integral operators. All informations are contained in the internal subrange Ω of the definition domain Ω_{def}, which is complementary to the external subrange, because only there the distributions do not have to disappear. This internal subrange Ω is supposed to be the glacier area under consideration. The oriented boundary surface from the internal subarea Ω to the external subarea Ω_{ext} belongs by definition to the internal subarea and is transverse and synchronous to[41] the model cone. The internal subarea Ω is compatible with the model cone and its boundary surface to the external subarea Ω_{ext}.[42]

---

[38] See Sect. 3.2.

[39] The domain of definition should be an open set. This requirement has mathematical reasons and is of little importance for the practical calculations.

[40] This outer subarea should also be an open set.

[41] See the definition of "transverse and synchronous" in Sect. 3.4.1.

[42] See the definition of compatibility in Sect. 3.4.1.

3.5.2 Integrations

The distributions χ are defined by the test results (3.37) obtained by testing them with so-called test functions τ. Test functions are all smooth functions declared on the domain of definition $\Omega_{\rm def}$ whose support is compact.[43] That the distributions χ vanish in the external domain $\Omega_{\rm ext}$ means that the test results vanish with all test functions τ whose carrier supp(τ) is in the external domain (3.38).

In order to apply the integral operators to the distributions, the associated integral operators $(\mathbf{a}\nabla)^{-1*}$ (3.40) and $\square_z^{-1*}$ (3.43) are introduced, respectively, which act only on the test functions τ. These associated integral operators arise from the original integral operators $(\mathbf{a}\nabla)^{-1}$ (3.39) and $\square_z^{-1}$ (3.42), respectively, by point mirroring their integration cones. The same computational rules apply to these associated integral operators as to the original integral operators,[44] in particular, the associated integral operators $(\mathbf{a}\nabla)^{-1*}$ and $\square_z^{-1*}$ are inverses of the corresponding differential operators $\mathbf{a}\nabla$ and $\square_z$.

Not only all smooth functions whose compact carrier lies in the domain of definition $\Omega_{\rm def}$ are used as test functions, but also all functions τ which can be generated by means of arbitrary products of associated integral operators. Indeed, with the functions τ generated in this way one can also test the distributions χ (3.37), although these functions τ in general no longer have a compact carrier.[45] The set of test functions defined in this way is closed under the associated integral operators, that is, these associated integral operators generate test functions again from test functions.

Using this set of test functions, the integral operators can also be applied to distributions and in this way integrals of distributions can be generated by rolling the integral operators from the distributions χ onto the test functions τ in the tests of these integrals and replacing them by their associated integral operators (3.44) and (3.45). The application of integral operators to distributions defined in this way is consistent with the already established application to ordinary functions χ (3.46)–(3.49).

The rules of calculation for the application of differential and integral operators to functions are also valid for the application to distributions.[46] Arbitrary polynomials of

[43] The carrier supp(τ) of a test function τ is the set of all points where τ does not vanish, extended by its cluster points. Compact means that the carrier is spatially bounded.

[44] See Sect. 3.2.

[45] The carrier of a function τ generated by associated integration from an integrand with compact carrier lies in the region irradiated by the compact carrier of the integrand through the model cones, i.e. covered by all rays of the model cones emanating from all points of the compact carrier. The infinitely long tail of this irradiated region extends into the external region $\Omega_{\rm ext}$ where all distributions vanish. Therefore, in all tests, this function τ can be represented by a substitute function with spatially restricted carrier, which differs from the original function τ only in the external region $\Omega_{\rm ext}$ where the distributions vanish.

[46] Justification: These computational rules given in Sect. 3.2 by (3.21)–(3.27) can be proved by "rolling over" the integral operators given by (3.44)–(3.45) from the distributions to the test functions.

differential and integral operators are defined, since these operators generate such distributions again from distributions which vanish in the external domain.

$$\int dV \cdot \tau \cdot \chi \tag{3.37}$$

$$\int dV \cdot \tau \cdot \chi = 0; \quad \text{supp}(\tau) \overset{\text{prec.}}{\subset} \Omega_{\text{ext}} \tag{3.38}$$

* * *

$$\left[(\mathbf{a}\nabla)^{-1}\chi\right](\mathbf{r}) \overset{\text{def.}}{=} -\int_0^{\infty} d\alpha \cdot \chi(\mathbf{r} + \alpha\mathbf{a}) \tag{3.39}$$

$$\left[(\mathbf{a}\nabla)^{-1*}\tau\right](\mathbf{r}) \overset{\text{def.}}{=} \int_{-\infty}^{0} d\alpha \cdot \tau(\mathbf{r} + \alpha\mathbf{a}) \tag{3.40}$$

$$G(\mathbf{r}' - \mathbf{r}) \overset{(3.20)}{=} \frac{1}{2\pi} \cdot \frac{\theta\left[(z' - z) - \sqrt{(x' - x)^2 + (y' - y)^2}\right]}{\sqrt{(z' - z)^2 - (x' - x)^2 - (y' - y)^2}} \tag{3.41}$$

$$\left[\Box_z^{-1}\chi\right](\mathbf{r}) \overset{\text{def}}{=} \int dV' \cdot G(\mathbf{r}' - \mathbf{r}) \cdot \chi(\mathbf{r}') \tag{3.42}$$

$$\left[\Box_z^{-1*}\tau\right](\mathbf{r}) \overset{\text{def}}{=} \int dV' \cdot G(\mathbf{r} - \mathbf{r}') \cdot \tau(\mathbf{r}') \tag{3.43}$$

* * *

$$(\mathbf{a}\nabla)^{-1}\chi : \quad \int dV \cdot \tau \cdot (\mathbf{a}\nabla)^{-1}\chi \overset{\text{def}}{=} -\int dV \cdot \left[(\mathbf{a}\nabla)^{-1*}\tau\right] \cdot \chi \tag{3.44}$$

$$\Box_z^{-1}\chi : \quad \int dV \cdot \tau \cdot \Box_z^{-1}\chi \overset{\text{def}}{=} \int dV \cdot \left[\Box_z^{-1*}\tau\right] \cdot \chi \tag{3.45}$$

* * *

$$0 = \int \mathrm{d}V \cdot \mathbf{a}\nabla\left\{(\mathbf{a}\nabla)^{-1*}\tau \cdot (\mathbf{a}\nabla)^{-1}\chi\right\}$$

$$\stackrel{\text{id.}}{=} \int \mathrm{d}V \cdot \tau \cdot (\mathbf{a}\nabla)^{-1}\chi + \int \mathrm{d}V \cdot \chi \cdot (\mathbf{a}\nabla)^{-1*}\tau \tag{3.46}$$

$$p \stackrel{\text{def}}{=} \Box_z^{-1*}\tau \tag{3.47}$$

$$q \stackrel{\text{def}}{=} \Box_z^{-1}\chi \tag{3.48}$$

$$\begin{aligned} 0 &= \int \mathrm{d}V \cdot \partial_z(\partial_z p \cdot q - p \cdot \partial_z q) \\ &- \int \mathrm{d}V \cdot \partial_x(\partial_x p \cdot q - p \cdot \partial_x q) \\ &- \int \mathrm{d}V \cdot \partial_y(\partial_y p \cdot q - p \cdot \partial_y q) \\ &\stackrel{\text{id.}}{=} \int \mathrm{d}V[q \cdot \Box_z p - p \cdot \Box_z q] \end{aligned}$$

$$\stackrel{\text{id.}}{=} \int \mathrm{d}V \cdot \tau \cdot \Box_z^{-1}\chi - \int \mathrm{d}V \cdot \chi \cdot \Box_z^{-1*}\tau \tag{3.49}$$

Forces and Torques on Surfaces

4

Abstract

The following transformations and terms are used to represent the forces and torques generated by a stress tensor field on oriented surfaces.

4.1 Gauss's Theorem

First, only those oriented surfaces Γ are considered whose surface normals **n** have everywhere non-positive z-components. If one applies Gauss's theorem to the so-called projection shadow $\omega_z(\Gamma)$, which is cast by the surface Γ in positive z-direction, then the force (4.1) on the oriented surface Γ can be written as difference of a volume integral over the projection shadow and of the force on the oriented[1] cylindrical shell surface Γ' of the projection shadow. The same applies to the torque (4.2). It is assumed that **S** is defined in all space and vanishes identically if one goes far enough in the positive z-direction so that the volume integrals over the infinitely long projection shadow $\omega_z(\Gamma)$ and the area integrals over its infinitely long lateral surface Γ' are finite.[2]

The area element dA (4.3) of the shell surface Γ' is given by the differentials of the z-coordinate and the arc length l on the lines of constant z. Thus, the force and torque on this shell surface Γ' can be converted into path integrals (4.6) and (4.7) over the oriented boundary curve $\partial\Gamma$ of the surface Γ. Consequently, the force on the oriented surface Γ is equal to the sum (4.8) of a volume integral over the projection shadow $\omega_z(\Gamma)$ of the surface

[1] The normal is directed outwards.

[2] See (2.10) and (2.11). The transformations (4.1)–(4.9) are identical transformations of expressions containing a matrix field **S** defined in the whole space, independent of whether this field satisfies balance conditions.

P. Halfar, *Stresses in glaciers*, https://doi.org/10.1007/978-3-662-66024-9_4

and a path integral over the oriented boundary curve $\partial\Gamma$ of the surface. The same applies to the torque (4.9).

These relations, so far restricted to oriented surfaces Γ with everywhere negative z-components of their surface normals, can be transferred to arbitrary oriented surfaces Γ by defining the projection shadows $\omega_z(\Gamma)$ of the oriented surfaces Γ as oriented additive entities and the volume integrals over the projection shadows as oriented additive functions of the oriented surfaces. This is carried out in the following.

—

$$\int_{\Gamma} \mathbf{Sn} \cdot \mathrm{d}A \overset{\text{id.}}{=} \int_{\omega_z(\Gamma)} \operatorname{div} \mathbf{S} \cdot \mathrm{d}V - \int_{\Gamma'} \mathbf{Sn} \cdot \mathrm{d}A \tag{4.1}$$

$$\int_{\Gamma} \mathbf{r} \times \mathbf{Sn} \cdot \mathrm{d}A \overset{\text{id.}}{=} \int_{\omega_z(\Gamma)} [\mathbf{r} \times \operatorname{div} \mathbf{S} + 2 \cdot \not{\mathbf{S}}] \cdot \mathrm{d}V - \int_{\Gamma'} \mathbf{r} \times \mathbf{Sn} \cdot \mathrm{d}A \tag{4.2}$$

* * *

$$\mathrm{d}A \overset{\text{id.}}{=} \mathrm{d}z \cdot \mathrm{d}l \tag{4.3}$$

$$\mathbf{n} \cdot \mathrm{d}l \overset{\text{id.}}{=} \mathbf{e}_z \times \mathrm{d}\mathbf{r} \overset{\text{id.}}{=} \not{\mathbf{e}}_z \cdot \mathrm{d}\mathbf{r} \tag{4.4}$$

$$\partial_z^{-1}(\not{\mathbf{r}}\mathbf{S}) \overset{\text{id.}}{=} \not{\mathbf{r}} \cdot \partial_z^{-1}\mathbf{S} - \partial_z^{-1}[\underbrace{(\partial_z \not{\mathbf{r}})}_{\not{\mathbf{e}}_z} \cdot \partial_z^{-1}\mathbf{S}] \tag{4.5}$$

$$\int_{\Gamma'} \mathbf{Sn} \cdot \mathrm{d}A \overset{\text{id.}}{=} \int_{\Gamma'} \mathbf{Sn} \cdot \mathrm{d}z \cdot \mathrm{d}l \overset{\text{id.}}{=} - \oint_{\partial\Gamma} \partial_z^{-1}\mathbf{S} \cdot \not{\mathbf{e}}_z \cdot \mathrm{d}\mathbf{r} \tag{4.6}$$

$$\int_{\Gamma'} \mathbf{r} \times \mathbf{Sn} \cdot \mathrm{d}A \overset{\text{id.}}{=} \int_{\Gamma'} \not{\mathbf{r}} \cdot \mathbf{Sn} \cdot \mathrm{d}z \cdot \mathrm{d}l \overset{\text{id.}}{=} - \oint_{\partial\Gamma} \partial_z^{-1}(\not{\mathbf{r}}\mathbf{S}) \cdot \not{\mathbf{e}}_z \cdot \mathrm{d}\mathbf{r}$$

$$\overset{\text{id.}}{=} \oint_{\partial\Gamma} \left(-\not{\mathbf{r}} \cdot \partial_z^{-1}\mathbf{S} + \not{\mathbf{e}}_z \cdot \partial_z^{-2}\mathbf{S} \right) \cdot \not{\mathbf{e}}_z \cdot \mathrm{d}\mathbf{r} \tag{4.7}$$

* * *

$$\int_{\Gamma} \mathbf{Sn} \cdot \mathrm{d}A \stackrel{\text{id.}}{=} \int_{\omega_z(\Gamma)} \operatorname{div} \mathbf{S} \cdot \mathrm{d}V + \oint_{\partial \Gamma} \partial_z^{-1} \mathbf{S} \cdot \not{\mathbf{e}}_z \cdot \mathrm{d}\boldsymbol{r} \tag{4.8}$$

$$\int_{\Gamma} \mathbf{r} \times \mathbf{Sn} \cdot \mathrm{d}A \stackrel{\text{id.}}{=} \int_{\omega_z(\Gamma)} [\mathbf{r} \times \operatorname{div} \mathbf{S} + 2 \cdot \not{\mathbf{S}}] \cdot \mathrm{d}V + \oint_{\partial \Gamma} \left(\not{\mathbf{r}} \cdot \partial_z^{-1} \mathbf{S} - \not{\mathbf{e}}_z \cdot \partial_z^{-2} \mathbf{S} \right) \cdot \not{\mathbf{e}}_z \cdot \mathrm{d}\mathbf{r} \tag{4.9}$$

4.2 Projection Shadow

The projection shadows $\omega_z(\Gamma)$ (Fig. 4.1) of oriented surfaces Γ cast in the positive z-direction are declared to be oriented and additive entities in the following way:

- If the orientation of the surface changes, the orientation of its shadow also reverses.
- If two oriented surfaces are joined to form a larger oriented surface,[3] their oriented shadows also merge, eliminating overlaps with different orientations.
- The shadows of surfaces whose surface normals have no positive z-components are positively oriented.

Because of their additivity, the projection shadows of any oriented surfaces can be created by unifying the oriented shadows of their surface elements, eliminating overlays with different orientations.

4.3 Oriented Volume Integrals

The oriented volume integral over an oriented projection shadow $\omega_z(\Gamma)$ is by definition equal to the ordinary integral, signed with the orientation sign of the projection shadow. If the projection shadow $\omega_z(\Gamma)$ of the oriented surface Γ consists of several regions with different orientations, as in Figure b of Fig. 4.1, then the oriented volume integral consists of the sum over the orientation-signed ordinary volume integrals over these regions. An oriented volume integral is an oriented and additive function of oriented surfaces Γ. The function value changes sign when the area orientation is reversed, and the function values of two oriented areas add together when these areas are combined to form a larger oriented area.

[3] A larger oriented surface is composed of two smaller oriented surfaces if these two smaller surfaces are obtained by cutting the larger surface.

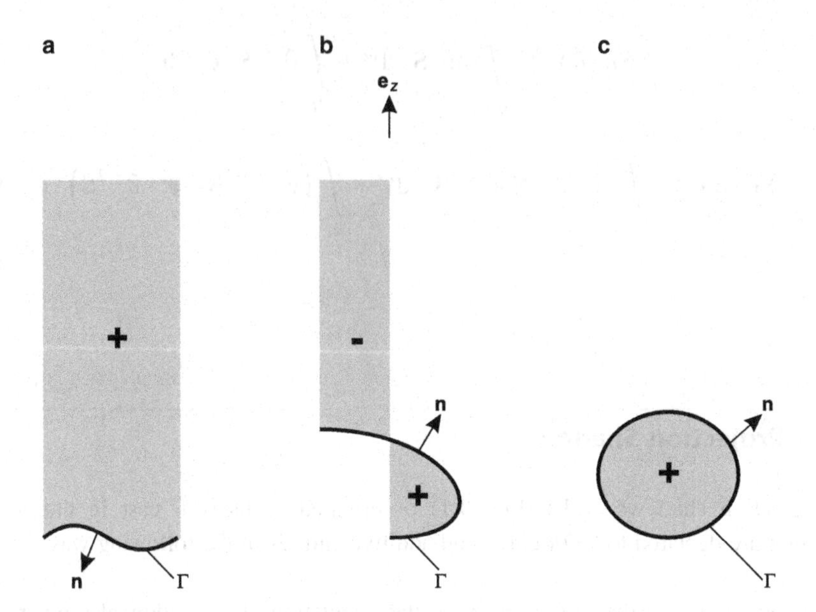

Fig. 4.1 Oriented projection shadows of oriented surfaces cast in the positive z-direction Γ. The figures show sections parallel to the z-direction through oriented surfaces Γ and through their projection shadows. (**a**) The oriented surface normals **n** have no positive z-components and the projection shadow of the oriented surface Γ is therefore positively oriented. (**b**) The oriented surface Γ has more complicated shape and its projection shadow consists of two parts with different orientation. (**c**) The projection shadow of the closed surface consists of the area enclosed by it and has positive orientation because the surface normals point outwards

4.4 Projection Masses and Moments

The expressions for the force (4.8) and the torque (4.9), which are generated by a stress tensor field **S** on an oriented surface Γ, are based on Gauss's theorem and are therefore mathematical identities for this stress tensor field. Physical relations arise from it by considering the balance conditions (2.14) and (2.15). Thus the force and the torque on the oriented surface Γ can be expressed by the so-called projection mass $m_z(\Gamma)$ (4.10) and the so-called projection moment $\mathbf{M}_z(\Gamma)$ (4.11) of this surface, respectively, and by path integrals over its oriented boundary $\partial\Gamma$ (4.12) and (4.13).

The projection mass $m_z(\Gamma)$ (4.10) of an oriented surface Γ is the oriented volume integral of the ice density ρ and is equal to the ordinary ice mass located in the oriented projection shadow of Γ, provided with the orientation sign of the shadow. Here the ice density ρ becomes a function defined in the whole space by being continued into the region outside the finite glacier area under consideration by the value zero, so that the projection masses are always defined and finite, even in the case of an infinitely extended projection shadow. If the shadow of the oriented surface Γ consists of several areas with different orientations

(Fig. 4.1b), then the projection mass $m_z(\Gamma)$ consists of the sum over the orientation-signed masses in these areas. In the special case of a closed oriented surface with outward pointing normals (Fig. 4.1c), the projection mass consists of the ordinary included mass. The same applies to the projection moments $\mathbf{M}_z(\Gamma)$ (4.11).

The projection masses $m_z(\Gamma)$ and projection moments $\mathbf{M}_z(\Gamma)$ thus defined are oriented and additive functions of oriented surfaces Γ. Their function value changes sign when the surface orientation ($\Gamma \rightarrow -\Gamma$) is reversed (4.14) and the function values of two oriented surfaces Γ_1 and Γ_2 add, (4.15) when these surfaces are joined to form a larger oriented surface $\Gamma_1 + \Gamma_2$. Because of their additivity, projection masses and projection moments of any oriented surfaces can also be calculated by integrating over the differential projection masses and projection moments of the oriented surface elements.

Projection masses and projection moments with respect to other projection directions can be defined accordingly. In the following still the projection masses and projection moments $m_x(\Gamma)$, $\mathbf{M}_x(\Gamma)$ and $m_y(\Gamma)$, $\mathbf{M}_y(\Gamma)$ occur due to projection in positive x- and y-direction respectively.

$$m_z(\Gamma) \overset{\text{def}}{=} \int_{\omega_z(\Gamma)} \rho \cdot \mathrm{d}V \tag{4.10}$$

$$\mathbf{M}_z(\Gamma) \overset{\text{def}}{=} \int_{\omega_z(\Gamma)} \mathbf{r} \cdot \rho \cdot \mathrm{d}V \tag{4.11}$$

* * *

$$\int_{\Gamma} \mathbf{Sn} \cdot \mathrm{d}A \overset{(4.8),(2.14)}{=} -m_z(\Gamma) \cdot \mathbf{g} + \oint_{\partial \Gamma} \partial_z^{-1} \mathbf{S} \cdot \not{\mathbf{e}}_z \cdot \mathrm{d}\mathbf{r} \tag{4.12}$$

$$\int_{\Gamma} \mathbf{r} \times \mathbf{Sn} \cdot \mathrm{d}A \overset{(4.9),(2.14)}{=} -\mathbf{M}_z(\Gamma) \times \mathbf{g} + \oint_{\partial \Gamma} \left(\not{\mathbf{r}} \cdot \partial_z^{-1} \mathbf{S} - \not{\mathbf{e}}_z \cdot \partial_z^{-2} \mathbf{S} \right) \cdot \not{\mathbf{e}}_z \cdot \mathrm{d}\mathbf{r} \tag{4.13}$$

* * *

$$m_z(\Gamma) = -m_z(-\Gamma); \quad \mathbf{M}_z(\Gamma) = -\mathbf{M}_z(-\Gamma) \tag{4.14}$$

$$m_z(\Gamma_1 + \Gamma_2) = m_z(\Gamma_1) + m_z(\Gamma_2); \quad \mathbf{M}_z(\Gamma_1 + \Gamma_2) = \mathbf{M}_z(\Gamma_1) + \mathbf{M}_z(\Gamma_2) \tag{4.15}$$

Fig. 4.1b), then the projection mass $m_q(\Gamma)$ consists of the sum over the orientation-signed masses in these areas. In the special case of a closed oriented surface with outward pointing normals (Fig. 4.1c), the projection mass consists of the ordinary included mass. The same applies to the projection moments $\mathbf{M}_q(\Gamma)$ (4.11).

The projection masses $m_q(\Gamma)$ and projection moments $\mathbf{M}_q(\Gamma)$ thus defined are oriented and additive functions of oriented surfaces Γ. Their function values change sign when the surface orientation is reversed (4.14) and the function values of two oriented surfaces Γ_1 and Γ_2 add (4.15) when these surfaces are joined to form a larger oriented surface (4.15). Because of their additivity, projection masses and projection moments of any oriented surfaces can also be determined by integrating over the infinitesimal projection masses and projection moments of the oriented surface elements.

Projection masses and projection moments with respect to other projection directions can be defined accordingly. In the following still the projection masses and projection moments m_q, $\mathbf{M}_q$ and m_{-q}, $\mathbf{M}_{-q}$ occur due to projection in positive and negative direction respectively:

5 Special Solutions of the Balance Conditions

Abstract

As a first step to construct the general solution of the balance and boundary conditions, two special solutions of only the balance conditions are computed.

As a first step to construct the general solution of the balance and boundary conditions, two special solutions of the balance conditions (2.14) and (2.15) are to be calculated.[1] Since all that matters here is to find some special solution, the two solutions can be constructed in such a way that their mathematical structure is simple: in each case three of their six independent components vanish and the three non-vanishing components consist of integer powers of the differential operators applied to the ice density ρ (5.1). Here the domain of definition of this function ρ is extended to the whole space by assigning it the value zero outside the considered finite range of glacier.[2] For the operator products and their integer powers simple calculation rules apply (5.2) and (5.3).[3]

$$\partial_x^l \cdot \partial_y^m \cdot \partial_z^n \cdot \rho; \quad l,m,n = \ldots -1,0,1 \ldots \tag{5.1}$$

[1] See footnote 8 in Chap. 2.

[2] The vectors $\mathbf{e}_x$, $\mathbf{e}_y$ and $\mathbf{e}_z$ are, according to the agreement in Sect. 3.2, vectors of the one-dimensional integration cones of the integral operators ∂_x^{-1}, ∂_y^{-1} and ∂_z^{-1}. According to Sect. 3.2, these integral operators and thus all integer powers of the differential operators are defined on the ice density ρ, since this function ρ vanishes outside the finite range of glacier under consideration.

[3] See (3.25) and (3.26).

P. Halfar, *Stresses in glaciers*, https://doi.org/10.1007/978-3-662-66024-9_5

$$\partial_i^l \cdot \partial_k^m = \partial_k^m \cdot \partial_i^l; \quad l,m = \ldots -1,0,1\ldots; \quad i,k = x,y,z \tag{5.2}$$

$$\partial_i^l \cdot \partial_i^m = \partial_i^{l+m}; \quad l,m = \ldots -1,0,1\ldots; \quad i = x,y,z \tag{5.3}$$

5.1 Disappearing *xx*, *xy* and *yy* Components

The stress tensor field $\mathbf{S}_b$ (5.4) is a everywhere defined special solution of the balance conditions (2.14) and (2.15) with vanishing *xx*-, *xy*- and *yy*-components.[4] Using the rules of calculation for the integer powers of the differential operators (5.2) and (5.3), one immediately sees that the balance conditions are satisfied by $\mathbf{S}_b$. The forces (4.12) and torques (4.13), which are generated by the stress tensor field $\mathbf{S}_b$ on oriented surfaces Γ, (5.8) and (5.9) depend on the projection masses $m_z(\Gamma)$ and projection moments $\mathbf{M}_z(\Gamma)$ of these surfaces, respectively and on path integrals over the oriented boundary curves $\partial\Gamma$ of these surfaces with the double integral $\partial_z^{-2}\rho$ of the ice density in the integrand.

With the z-axis oriented vertically, the zz-component is the only non-vanishing component (5.10) and the forces (5.11) and torques (5.12) on oriented surfaces Γ depend in this case only on the projection masses $m_z(\Gamma)$ or projection moments $\mathbf{M}_z(\Gamma)$ of these surfaces.

—

$$\mathbf{S}_b = \mathbf{S}_b^T \stackrel{\text{def.}}{=} \begin{bmatrix} 0 & 0 & -g_x \cdot \partial_z^{-1} \\ 0 & 0 & -g_y \cdot \partial_z^{-1} \\ * & * & (g_x \cdot \partial_x + g_y \cdot \partial_y)\partial_z^{-2} - g_z \cdot \partial_z^{-1} \end{bmatrix} \cdot \rho$$
$$\stackrel{\text{id.}}{=} \left[-(\mathbf{g}\mathbf{e}_z^T + \mathbf{e}_z\mathbf{g}^T) \cdot \partial_z^{-1} + \mathbf{e}_z\mathbf{e}_z^T \cdot \partial_z^{-2} \cdot (\mathbf{g}^T\nabla) \right] \cdot \rho \tag{5.4}$$

* * *

$m = \ldots -1,0,1\ldots:$

[4] The designation $\mathbf{S}_b$ was chosen to be consistent with the designations in Sect. 8.2. This stress tensor field $\mathbf{S}_b$ depends on the orientation of the z-axis, but it is invariant to rotations of the coordinate system around the z-axis, because in this case the tensor components of $\mathbf{S}_b$are transformed accordingly. Thus, there are actually an infinite number of stress tensor fields $\mathbf{S}_b$, namely one for each orientation of the z-axis.

$$\partial_z^m \mathbf{S}_b \cdot \not{\mathbf{e}}_z \overset{\text{id.}}{=} \mathbf{e}_z (\mathbf{e}_z \times \mathbf{g})^{\mathrm{T}} \cdot \partial_z^{m-1} \rho \tag{5.5}$$

$$-\not{\mathbf{r}} \cdot \partial_z^m \mathbf{S}_b \cdot \not{\mathbf{e}}_z \overset{\text{id.}}{=} (\mathbf{e}_z \times \mathbf{r})(\mathbf{e}_z \times \mathbf{g})^{\mathrm{T}} \cdot \partial_z^{m-1} \rho \tag{5.6}$$

$$\not{\mathbf{e}}_z \cdot \partial_z^m \mathbf{S}_b \cdot \not{\mathbf{e}}_z \overset{\text{id.}}{=} 0 \tag{5.7}$$

* * *

$$\int_\Gamma \mathbf{S}_b \mathbf{n} \cdot dA \overset{(4.12)}{=} -m_z(\Gamma) \cdot \mathbf{g} + \mathbf{e}_z \cdot \oint_{\partial\Gamma} \partial_z^{-2} \rho \cdot (\mathbf{e}_z \times \mathbf{g}) \cdot d\mathbf{r} \tag{5.8}$$

$$\int_\Gamma \mathbf{r} \times \mathbf{S}_b \mathbf{n} \cdot dA \overset{(4.13)}{=} -\mathbf{M}_z(\Gamma) \times \mathbf{g} - \mathbf{e}_z \times \oint_{\partial\Gamma} \partial_z^{-2} \rho \cdot \mathbf{r} \cdot [(\mathbf{e}_z \times \mathbf{g}) \cdot d\mathbf{r}] \tag{5.9}$$

* * *

$$\mathbf{g} \overset{\text{prec.}}{=} g_z \cdot \mathbf{e}_z :$$

$$\mathbf{S}_b = \mathbf{S}_b^{\mathrm{T}} \overset{\text{def}}{=} \begin{bmatrix} 0 & 0 & 0 \\ 0 & 0 & 0 \\ 0 & 0 & -g_z \cdot \partial_z^{-1} \rho \end{bmatrix} \overset{\text{id.}}{=} -g_z \cdot \partial_z^{-1} \rho \cdot \mathbf{e}_z \mathbf{e}_z^{\mathrm{T}} \tag{5.10}$$

$$\int_\Gamma \mathbf{S}_b \mathbf{n} \cdot dA \overset{\text{id.}}{=} -m_z(\Gamma) \cdot \mathbf{g} \tag{5.11}$$

$$\int_\Gamma \mathbf{r} \times \mathbf{S}_b \mathbf{n} \cdot dA \overset{\text{id.}}{=} -\mathbf{M}_z(\Gamma) \times \mathbf{g} \tag{5.12}$$

5.2 Disappearing Non-diagonal Components

The stress tensor field $\mathbf{S}_e$ (5.13) is a everywhere defined special solution of the balance conditions (2.14) and (2.15) and its components outside the diagonal vanish.[5] As in the case of the stress tensor field $\mathbf{S}_b$, it can be proved with the help of the simple calculation rules (5.2) and (5.3) for the integer powers of the differential operators that the balance

[5]The designation was $\mathbf{S}_e$ chosen to be consistent with the designations in Sect. 8.2. The diagonal stress tensor field $\mathbf{S}_e$ generally depends on the orientation of the coordinate system. Only in the trivial special case of horizontally homogeneous ice density and horizontal, free ice surface $\mathbf{S}_e$ is independent of the orientation of the coordinate system and agrees with the trivial stagnant solution, where the base and integration cone vectors $\mathbf{e}_i$should point upwards.

conditions are fulfilled by $\mathbf{S}_e$. This time the forces and torques on oriented surfaces Γ shall not be calculated by the above described method (4.12) and (4.13), but by a comparison with the forces (5.11) and torques (5.12) for the stress tensor field $\mathbf{S}_b$ (5.10). Thus, the force (5.14) and torque (5.15) generated by the stress tensor field $\mathbf{S}_e$ on an oriented surface Γ can be expressed by the three projection masses $m_i(\Gamma)$ and the three projection moments $\mathbf{M}_i(\Gamma)$ $(i = x, y, z)$ of that surface, respectively.

$$\mathbf{S}_e = \mathbf{S}_e^{\mathrm{T}} \overset{\text{def}}{=} -\begin{bmatrix} g_x \cdot \partial_x^{-1} & 0 & 0 \\ 0 & g_y \cdot \partial_y^{-1} & 0 \\ 0 & 0 & g_z \cdot \partial_z^{-1} \end{bmatrix} \cdot \rho \tag{5.13}$$

$$\int_\Gamma \mathbf{S}_e \mathbf{n} \cdot dA \overset{\text{id.}}{=} -g_x \cdot m_x(\Gamma) \cdot \mathbf{e}_x - g_y \cdot m_y(\Gamma) \cdot \mathbf{e}_y - g_z \cdot m_z(\Gamma) \cdot \mathbf{e}_z \tag{5.14}$$

$$\int_\Gamma \mathbf{r} \times \mathbf{S}_e \mathbf{n} \cdot dA \overset{\text{id.}}{=} -g_x \cdot \mathbf{M}_x(\Gamma) \times \mathbf{e}_x - g_y \cdot \mathbf{M}_y(\Gamma) \times \mathbf{e}_y - g_z \cdot \mathbf{M}_z(\Gamma) \times \mathbf{e}_z \tag{5.15}$$

6 Weightless Stress Tensor Fields

Abstract

In this chapter, the “weightless stress tensor fields” are presented. These “weightless stress tensor fields” form the general solution of the homogenized balance conditions, in which the ice density is considered as a formal parameter and is set to zero.

The computation of the general solution of the balance and boundary conditions (2.14)–(2.16) could be reduced to the simpler problem of computing the general weightless solution of the balance and boundary conditions (2.21)–(2.23) for weightless stress tensor fields by subtracting a special solution.[1,2]

As a first step towards the construction of this general weightless solution of the balance and boundary conditions (2.21)–(2.23), this chapter constructs the general weightless solution of only the balance conditions (2.21)–(2.22) using a long known procedure [2, pp. 53–57]. By definition, this general solution consists of all weightless stress tensor fields.

6.1 Construction

The weightless stress tensor fields $\mathbf{T}$ can be characterized by the fact that they do not generate forces (2.25) and torques (2.26) on closed surfaces. This means that the rows of matrix fields $\mathbf{T}$ and $\boldsymbol{r} \cdot \mathbf{T}$ have vanishing divergences and can be written as transposed

[1] Such special solutions are, for example, the stress tensor fields $\mathbf{S_b}$ (5.4) and $\mathbf{S_e}$ (5.13).

[2] S. Chap. 2.

P. Halfar, *Stresses in glaciers*, https://doi.org/10.1007/978-3-662-66024-9_6

rotations (6.1) and (6.2) of matrix fields **B** respectively **C**.[3] By transformations (6.3)–(6.8), all weightless stress tensor fields **T** can be obtained from matrix fields **A** by second order derivatives (6.9). The matrix fields **A** are called stress functions [2, p. 54] or **A**-fields. This form of representability of weightless stress tensor fields **T** by stress functions **A** is not only a necessary condition, but also a sufficient one in the sense that any **A**-field defines matrix fields **B** (6.8), **T** (6.9) and **C** (6.10) satisfying the requirements (6.1) and (6.2) imposed on them.

All fields **T** (6.9) defined in this way form the totality of the weightless stress tensor fields. Thus there is a simple representation of these stress tensor fields and this representation depends only on the symmetrical parts $\mathbf{A}_{+}$ of the **A**-fields.[4] These weightless stress tensor fields **T** generate forces (6.12) and torques (6.13) on oriented surfaces Γ, which are given according to Stokes' theorem by path integrals of the **B**- (6.8) or **C**-fields (6.10) over the oriented edge ∂Γ of this surface, whereby also here only the symmetrical parts of the **A**-fields play a role.

If matrix fields **B** and **C** as described above are defined by a matrix field **A** (6.14) and (6.15), they shall be called "descended from **A**" or "successors of **A**". Then they must satisfy the descent condition (6.16). This descent-condition is not only necessary, but also sufficient for these descent-relations. For if there are two matrix fields **B** and **C** which satisfy this descent condition (6.17), then there is also a matrix field **A** (6.18) from which they descend. Therefore, the weightless stress tensor fields **T** can also be obtained from all pairs of matrix fields **B** and **C** which satisfy the descent condition (6.17), which was the starting point for constructing the weightless stress tensor fields **T** (6.1) and (6.2).

—

$$\mathbf{T} = (\mathrm{rot}\,\mathbf{B})^T \tag{6.1}$$

$$\not{r} \cdot \mathbf{T} = (\mathrm{rot}\mathbf{C})^{\mathrm{T}} \tag{6.2}$$

* * *

$$\not{r} \cdot (\mathrm{rot}\mathbf{B})^{\mathrm{T}} \overset{\mathrm{id.}}{=} [\mathrm{rot}(\not{r} \cdot \mathbf{B})]^{\mathrm{T}} + \mathbf{B}^{\mathrm{T}} - \mathrm{trace}(\mathbf{B}) \cdot 1 \tag{6.3}$$

[3] The rotation rot **B** of a matrix field is defined by first transposing **B** and then replacing each of the three columns of this transposed matrix field with their rotation. (See (13.9) and [2, p. 11].) Thus the rows of the matrix field $(\mathrm{rot}\,\mathbf{B})^T$ consist of the rotations of the rows of **B**, which gives the desired representation of the rows of as **T** rotations. Analogous is true for $\mathbf{r} \cdot \mathbf{T}$.

[4] The weightless stress tensor field **T** is the dual rotation of the symmetric tensor field $\mathbf{A}_{+}$, where the dual rotation of a matrix field is given by (13.24). Gurtin [2, p. 54, 57] calls this form of a weightless stress tensor field a "solution of Beltrami".

$$\mathbf{B}^{\mathrm{T}} - \mathrm{trace}(\mathbf{B}) \cdot 1 = [\mathrm{rot}(\underbrace{\underbrace{\mathbf{C} - \not{\mathbf{r}} \cdot \mathbf{B}}\,\underbrace{\quad}}_{\mathbf{A}})]^{\mathrm{T}} \tag{6.4}$$

$$\mathbf{A} \overset{\mathrm{def}}{=} \mathbf{C} - \not{\mathbf{r}} \cdot \mathbf{B} \tag{6.5}$$

$$\mathbf{B}^{T} - \mathrm{trace}(\mathbf{B}) \cdot \mathbf{1} = (\mathrm{rot}\,\mathbf{A})^{\mathrm{T}} \tag{6.6}$$

$$\mathrm{trace}(\mathbf{B}) = -\frac{1}{2} \cdot \mathrm{trace}(\mathrm{rot}\,\mathbf{A}) \tag{6.7}$$

* * *

$$\mathbf{B} = \mathrm{rot}\mathbf{A} - \frac{1}{2} \cdot \mathrm{trace}(\mathrm{rot}\mathbf{A}) \cdot 1 \overset{\mathrm{id.}}{=} \mathrm{rot}\mathbf{A}_{+} - \mathrm{grad}\not{\mathbf{A}}. \tag{6.8}$$

$$\mathbf{T} = (\mathrm{rot}\,\mathbf{B})^{\mathrm{T}} = \mathrm{rot}\ \mathrm{rot}\mathbf{A}_{+} \overset{\mathrm{id.}}{=} [\mathrm{rot}\ \mathrm{rot}\mathbf{A}]_{+} \overset{\mathrm{id.}}{=} \mathbf{T}^{\mathrm{T}} \tag{6.9}$$

$$\mathbf{C} = \not{\mathbf{r}} \cdot \mathbf{B} + \mathbf{A} = \not{\mathbf{r}} \cdot \left[\mathrm{rot}\mathbf{A} - \frac{1}{2} \cdot \mathrm{trace}(\mathrm{rot}\mathbf{A}) \cdot 1\right] + \mathbf{A}$$

$$\overset{\mathrm{id.}}{=} \not{\mathbf{r}} \cdot \mathrm{rot}\mathbf{A}_{+} + \mathbf{A}_{+} - \mathrm{grad}(\not{\mathbf{r}} \cdot \not{\mathbf{A}}) \tag{6.10}$$

$$\not{\mathbf{r}}\mathbf{T} = (\mathrm{rot}\mathbf{C})^{\mathrm{T}} = [\mathrm{rot}(\not{\mathbf{r}} \cdot \mathrm{rot}\mathbf{A}_{+} + \mathbf{A}_{+})]^{\mathrm{T}} \tag{6.11}$$

* * *

$$\int_{\Gamma} \mathbf{Tn} \cdot \mathrm{d}A = \oint_{\partial\Gamma} \mathbf{B} \cdot \mathrm{d}\mathbf{r} = \oint_{\partial\Gamma} \mathrm{rot}\mathbf{A}_{+} \cdot \mathrm{d}\mathbf{r} \tag{6.12}$$

$$\int_{\Gamma} \not{\mathbf{r}}\mathbf{Tn} \cdot \mathrm{d}A = \oint_{\partial\Gamma} \mathbf{C} \cdot \mathrm{d}\mathbf{r} = \oint_{\partial\Gamma} (\not{\mathbf{r}} \cdot \mathrm{rot}\mathbf{A}_{+} + \mathbf{A}_{+}) \cdot \mathrm{d}\mathbf{r} \tag{6.13}$$

* * *

$$\mathbf{B} \overset{\mathrm{def.}}{=} \mathrm{rot}\,\mathbf{A} - \frac{1}{2} \cdot \mathrm{trace}(\mathrm{rot}\,\mathbf{A}) \cdot \mathbf{1} \tag{6.14}$$

$$\mathbf{C} \overset{\mathrm{def}}{=} \not{\mathbf{r}} \cdot \mathbf{B} + \mathbf{A} \tag{6.15}$$

$$(\mathrm{rot}\mathbf{C})^{\mathrm{T}} \overset{(13.22)}{=} \not{r} \cdot (\mathrm{rot}\mathbf{B})^{\mathrm{T}} \tag{6.16}$$

* * *

$$(\mathrm{rot}\mathbf{C})^{\mathrm{T}} \overset{\text{prec.}}{=} \not{r} \cdot (\mathrm{rot}\mathbf{B})^{\mathrm{T}} \tag{6.17}$$

$$\mathbf{A} \overset{\text{def}}{=} \mathbf{C} - \not{r} \cdot \mathbf{B} \tag{6.18}$$

$$\mathbf{B} \overset{(13.22)}{=} \mathrm{rot}\,\mathbf{A} - \frac{1}{2} \cdot \mathrm{trace}(\mathrm{rot}\,\mathbf{A}) \cdot \mathbf{1} \tag{6.19}$$

6.2 Redundancies and Normalizations

6.2.1 Redundancy Functions

Two different **A**-fields can lead to the same weightless stress tensor field **T** (6.9). This is exactly the case if the two **A**-fields differ by a so-called redundancy function $\mathbf{A}^{\bullet}$, which is characterized by the fact that its **T**-field $\mathbf{T}^{\bullet}$ vanishes. Arbitrary linear combinations of redundancy functions are also redundancy functions. Redundancy functions $\mathbf{A}^{\bullet}$ produce further **A**-fields from every **A**-field by addition, but these fields are redundant, because they have the same **T**-field. But only this **T**-field is important.

All redundancy functions $\mathbf{A}^{\bullet}$ (6.20) can be represented by any two vector fields **u** and **v**.[5] However, these redundancy functions can also be characterized by the fact that their symmetric parts $\mathbf{A}^{\bullet}_{+}$ (6.21) are symmetric gradients of arbitrary vector fields and their antisymmetric parts $\mathbf{A}^{\bullet}_{-}$ (6.22) are arbitrary. Their **B**- and **C**-fields $\mathbf{B}^{\bullet}$ (6.23) and $\mathbf{C}^{\bullet}$ (6.24), respectively, are gradient fields whose integrals over the closed boundary curves of arbitrary surfaces yield vanishing forces (6.12) and torques (6.13) on these surfaces, which is equivalent to the stress tensor fields $\mathbf{T}^{\bullet}$ (6.25) vanishing.

—

$$\mathbf{A}^{\bullet} = \not{\mathbf{u}} + \mathrm{grad}\mathbf{v} \tag{6.20}$$

$$\mathbf{A}^{\bullet}_{+} = \frac{1}{2}\left[\mathrm{grad}\,\mathbf{v} + (\mathrm{grad}\,\mathbf{v})^{\mathrm{T}}\right] \tag{6.21}$$

[5] See Sect. 14.1. $\not{\mathbf{u}}$ denotes the skew-symmetric tensor field associated with the vector field **u**.

$$\mathbf{A}_{-}^{\bullet} = \frac{1}{2}\left[\mathrm{grad}\mathbf{v} - (\mathrm{grad}\mathbf{v})^{\mathrm{T}}\right] + \not{\mathbf{u}} \tag{6.22}$$

$$\mathbf{B}^{\bullet} = \mathrm{rot}\mathbf{A}^{\bullet} - \frac{1}{2} \cdot \mathrm{trace}(\mathrm{rot}\mathbf{A}^{\bullet}) \cdot 1 = -\,\mathrm{grad}\,\mathbf{u} \tag{6.23}$$

$$\mathbf{C}^{\bullet} = \mathbf{A}^{\bullet} + \not{\mathbf{r}}\mathbf{B}^{\bullet} = \mathrm{grad}(\mathbf{v} - \not{\mathbf{r}}\mathbf{u}) = \mathrm{grad}(\mathbf{v} - \mathbf{r} \times \mathbf{u}) \tag{6.24}$$

$$\mathbf{T}^{\bullet} = (\mathrm{rot}\,\mathbf{B}^{\bullet})^{\mathrm{T}} = \mathbf{0} \tag{6.25}$$

6.2.2 Normalizations

Redundancies can be reduced by normalizing the **A**-fields. For the representation of all weightless stress tensor fields **T** then not all **A**-fields are used, but only those which fulfill a certain normalization condition, whereby the set of these normalized **A**-fields should still be large enough to generate all weightless stress tensor fields **T**. Thus, normalization should be admissible insofar as no weightless stress tensor fields **T** are lost. To guarantee this admissibility, it is sufficient to prove that each **A**-field can be transformed into a normalized **A**-field by adding a redundancy function.[6]

A possible normalization of **A**-fields admits only symmetric **A**-fields as normalized **A**-fields. This normalization is admissible, because every other **A**-field can be transformed into a symmetric **A**-field by subtraction of a redundancy function, namely its antisymmetric part $\mathbf{A}_{-}$. Gurtin [2, p. 54] calls this representation of all weightless stress tensor fields **T** (6.9) by symmetric-matrix fields **A** a "solution of Beltrami".

Even among the symmetric **A**-fields there are still redundancies which can be reduced by normalization. Various normalizations of the symmetric **A**-fields are possible, each consisting of only three of their six independent components not vanishing. Each choice of three such components represents a possible normalization, except for the cases where the three non-vanishing components are all in the same matrix row or column.[7] Five such normalizations are given below, that is, five combinations of three nonvanishing matrix elements among the six independent matrix elements of a symmetric matrix **A**. These normalizations are denoted by the indices xx-yy-zz, etc., of their three nonvanishing components. Each of these five normalizations belongs to one of five normalization types "A" to "E" defined below. The unspecified normalizations of a type are not essentially different from the specified normalization, but arise from it by renaming the location coordinates[8]:

[6] Or subtraction, which is equivalent to adding a redundancy function with the opposite sign.

[7] See Sect. 14.2.

[8] These renamings are equivalent to synchronous swaps of rows and columns of the matrix **A**.

A. Normalization type: three diagonal elements. Number of Normalizations of this type: 1

$$\mathrm{xx-yy-zz-normalization}: \mathbf{A} = \mathbf{A}^{\mathrm{T}} = \begin{bmatrix} * & 0 & 0 \\ 0 & * & 0 \\ 0 & 0 & * \end{bmatrix}$$

B. Normalization type: two diagonal elements and one in their intersection field. Number of Normalizations of this type: 3

$$\mathrm{xx-yy-xy-normalization}: \mathbf{A} = \mathbf{A}^{\mathrm{T}} = \begin{bmatrix} * & * & 0 \\ * & * & 0 \\ 0 & 0 & 0 \end{bmatrix}$$

C. Normalization type: two diagonal elements and one not in their intersection field. Number of normalizations of this type: 6

$$\mathrm{xx-yy-xz-normalization}: \mathbf{A} = \mathbf{A}^{\mathrm{T}} = \begin{bmatrix} * & 0 & * \\ 0 & * & 0 \\ * & 0 & 0 \end{bmatrix}$$

D. Normalization type: one diagonal element. Number of Normalizations of this type: 6

$$\mathrm{xx-xy-yz-normalization}: \mathbf{A} = \mathbf{A}^{\mathrm{T}} = \begin{bmatrix} * & * & 0 \\ * & 0 & * \\ 0 & * & 0 \end{bmatrix}$$

E. Normalization type: No diagonal element. Number of Normalizations of this type: 1

$$\mathrm{xy-yz-xz-normalization}: \mathbf{A} = \mathbf{A}^{\mathrm{T}} = \begin{bmatrix} 0 & * & * \\ * & 0 & * \\ * & * & 0 \end{bmatrix}$$

So there are 17 normalizations in total. These are the 20 combinatorial possibilities of choosing three out of six independent matrix elements of a symmetric matrix, minus the three “forbidden” cases in which the three matrix elements are each in a row or column.

All weightless stress tensor fields **T** (6.9) can thus be represented by **A**-fields which have one of these normalizations. Gurtin [2, p. 54, 55] calls the representation with the normalization xx-yy-zz “solution of Maxwell” and the representation with the normalization xy-yz-xz “solution of Morera”.

Part II

The General Solution

7 Weightless Stress Tensor Fields with Boundary Conditions

Abstract

In this chapter the "weightless stress tensor fields" with given boundary stresses are constructed by selecting among all "weightless stress tensor fields" those which have these boundary stresses. Since the "weightless stress tensor fields" are given as second derivatives of arbitrary matrix fields (stress functions), this selection is made by identifying suitable stress functions. This identification is done by boundary conditions for the boundary values of the stress functions and their first derivatives. If the boundary surface on which the boundary stresses are given contains a not simply connected part, then the part of the closed boundary of the glacier area under consideration on which no boundary stresses are given consists of several connected parts. The unknown forces and torques on these parts – except one of them – appear as parameters in the general solution.

In this chapter the weightless stress tensor fields with given boundary stresses are constructed. These stress tensor fields form the general weightless solution of the balance and boundary conditions (2.21)–(2.23) for weightless stress tensor fields. Thus, one has also solved the main task of this paper to construct the general solution of the balance and boundary conditions (2.14)–(2.16), since one can trace this main task back to the construction of the general weightless solution by subtracting an already known special solution of only the balance conditions.[1]

The weightless stress tensor fields with given boundary stresses are constructed by selecting among all weightless stress tensor fields $\mathbf{T}$ – these are already known – those

[1] Such special solutions are, for example, the stress tensor fields $\mathbf{S}_b$ (5.4) and $\mathbf{S}_e$ (5.13).

P. Halfar, *Stresses in glaciers*, https://doi.org/10.1007/978-3-662-66024-9_7

which satisfy the boundary condition (2.23). Since the weightless stress tensor fields **T** are given by second derivatives of arbitrary matrix fields **A**, this selection is made by identifying suitable **A**-fields. This identification is done by boundary conditions for the boundary values of the **A**-fields and their first derivatives.

7.1 Terms

The following terms play a role in the discussion of weightless stress tensor fields with given boundary stresses:

- **A**-∂_n**A**-boundary field
- B-C-boundary field
- **A**-solution set of a **A**-∂_n**A**-boundary field
- **T**-solution set of a **A**-∂_n**A**- boundary field and its **A**-solution set
- Marginal distribution of forces and torques

A-∂_nA- Boundary Fields, B-C- Boundary Fields, A- and T-Solution Sets
A **A**-∂_n**A** – boundary field describes the boundary values $\mathbf{A}_\Sigma$ and $\partial_n\mathbf{A}$ of a **A**-field and its normal derivatives on the boundary surface Σ. A **B**-**C**-boundary field consists of the boundary fields $\mathbf{B}_\Sigma$ and $\mathbf{C}_\Sigma$ of the fields **B** (6.8) and **C** (6.10) of a **A**-field. The **A**-solution set of a **A**-∂_n**A** – boundary field consists of all **A**-fields which together with their normal derivatives on the boundary surface Σ take the corresponding values of this **A**-∂_n**A**-boundary field and the **T**-solution set consists of all **T**-fields (6.9) of these **A**-fields.

Marginal Distributions of Forces and Torques
Marginal distribution of forces and torques means the totality of forces and torques, which the considered weightless stress tensor field **T** generates on the boundary surface Σ of given boundary stresses and on the connected components of their complement Λ. This complement Λ consists of the part of the simply closed boundary of ∂Ω the considered glacier area, on which no boundary conditions are set. The connected components of this complement Λ are denoted by $\Lambda_0, \ldots, \Lambda_n$ Possible configurations depend on the topological structure of the boundary surface Σ (Fig. 7.1). These configurations can be divided into four groups of cases:

- In the simplest case, Λ and Σ are simply connected and thus each have only a single separate connected component. In this case $\Lambda_0 = \Lambda$. (Fig. 7.1a)
- If Σ is disconnected and consists only of separate, simply connected parts, then Λ is multiply connected, $\Lambda_0 = \Lambda$. (Fig. 7.1b)

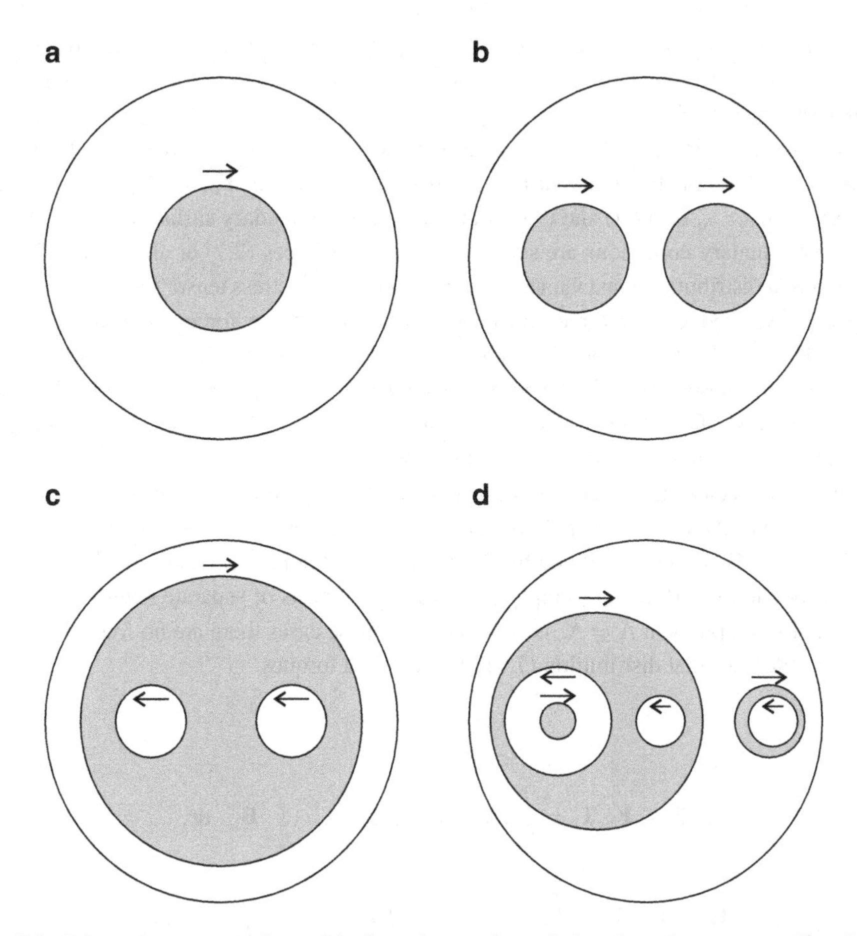

Fig. 7.1 Schematic representations of the boundary surface of given boundary stresses. The glacier area under consideration is schematically represented as a sphere. On the surface of the sphere lie the shaded boundary surface of given boundary stresses and the white surface of unknown boundary stresses. The arrows denote the orientations on the boundary curves of the white surface. The shaded and white surfaces are (**a**) connected and connected, respectively, (**b**) disconnected and connected, respectively, (**c**) connected and disconnected, respectively, (**d**) disconnected and disconnected, respectively

- If Σ is multiple connected, Λ consists of several separate, simple connected parts $\Lambda_0, \ldots, \Lambda_n$. (Fig. 7.1c)
- If Λ and Σ are disconnected, Λ consists of several separate connected parts $\Lambda_0, \ldots, \Lambda_n$, some of which are multiply connected. (Fig. 7.1d)

In all cases, the edges $\partial\Lambda_0, \ldots, \partial\Lambda_n$ of the separate, connected components of Λ each consist of one or more closed curves.[2] These boundary curves are at the same time boundary curves of Σ.

A marginal distribution of forces and torques thus consists of the force $\mathbf{F}_\Sigma$ (7.1) and the torque $\mathbf{G}_\Sigma$ (7.2) on the boundary surface Σ[3] as well as the forces $\mathbf{F}_0, \ldots, \mathbf{F}_n$ (7.3) and (7.5) and torques $\mathbf{G}_0, \ldots, \mathbf{G}_n$ (7.4) and (7.6) on the connected boundary surfaces $\Lambda_0, \ldots, \Lambda_n$, on which no boundary conditions are set. The sums of all forces (7.7) or all torques (7.8) of this marginal distribution must vanish, because a weightless stress tensor field on the closed boundary $\partial\Omega$ produces no force and no torque, so that these forces or torques are not independent. On the other hand, the forces $\mathbf{F}_1, \ldots, \mathbf{F}_n$ (7.5) or torques $\mathbf{G}_1, \ldots, \mathbf{G}_n$ (7.6) are independent because they can be specified arbitrarily within the general solution. They are free parameters of the general solution and describe the freely selectable marginal distributions of forces and torques on the connected boundary surfaces $\Lambda_1, \ldots, \Lambda_n$, i.e. on all connected boundary surfaces on which no boundary conditions are set, with the exception of the one connected boundary surface Λ_0. On this surface the force $\mathbf{F}_0$ (7.3) and the torque $\mathbf{G}_0$ (7.4) are defined by the balance conditions (7.7) and (7.8).[4]

If the boundary surface Σ is simply connected or consists of separate simply connected parts, the its complement $\Lambda = \Lambda_0$ is connected. In these cases there are no free parameters and only one marginal distribution (7.9) of forces and torques.

—

$$\mathbf{F}_\Sigma = \mathbf{F}_\Sigma[\mathbf{T}] = \int_\Sigma \mathbf{Tn} \cdot dA = \int_\Sigma \mathbf{t} \cdot dA = \oint_{\partial\Sigma} \mathbf{B}_\Sigma \cdot d\boldsymbol{r} \tag{7.1}$$

$$\mathbf{G}_\Sigma = \mathbf{G}_\Sigma[\mathbf{T}] = \int_\Sigma \mathbf{r} \times \mathbf{Tn} \cdot dA = \int_\Sigma \mathbf{r} \times \boldsymbol{t} \cdot dA = \oint_{\partial\Sigma} \mathbf{C}_\Sigma \cdot d\mathbf{r} \tag{7.2}$$

$$\mathbf{F}_0 = \mathbf{F}_0[\mathbf{T}] = \int_{\Lambda_0} \mathbf{Tn} \cdot dA = \oint_{\partial\Lambda_0} \mathbf{B}_\Sigma \cdot d\mathbf{r} \tag{7.3}$$

$$\mathbf{G}_0 = \mathbf{G}_0[\mathbf{T}] = \int_{\Lambda_0} \mathbf{r} \times \mathbf{Tn} \cdot dA = \oint_{\partial\Lambda_0} \mathbf{C}_\Sigma \cdot d\mathbf{r} \tag{7.4}$$

[2] The orientations of these curves are shown by arrows in Fig. 7.1. They result from the direction of rotation on $\partial\Omega$, which is counterclockwise when looking from the outside, and are determined by the outward pointing normal vectors on $\partial\Omega$.

[3] $\mathbf{F}_\Sigma$ and $\mathbf{G}_\Sigma$ are known, since they are defined by the given boundary stresses $\mathbf{t}$.

[4] In principle, it does not matter which of the connected boundary surfaces on which no boundary conditions are set is designated by Λ_0.

$$\mathbf{F}_\nu = \mathbf{F}_\nu[\mathbf{T}] = \int_{\Lambda_\nu} \mathbf{Tn} \cdot dA = \oint_{\partial\Lambda_\nu} \mathbf{B}_\Sigma \cdot \mathbf{dr}; \quad \nu = 1, \ldots, n \tag{7.5}$$

$$\mathbf{G}_\nu = \mathbf{G}_\nu[\mathbf{T}] = \int_{\Lambda_\nu} \mathbf{r} \times \mathbf{Tn} \cdot dA = \oint_{\partial\Lambda_\nu} \mathbf{C}_\Sigma \cdot \mathbf{dr}; \quad \nu = 1, \ldots, n \tag{7.6}$$

$$\mathbf{F}_\Sigma + \mathbf{F}_0 \cdots + \mathbf{F}_n = \mathbf{0} \tag{7.7}$$

$$\mathbf{G}_\Sigma + \mathbf{G}_0 \cdots + \mathbf{G}_n = \mathbf{0} \tag{7.8}$$

* * *

$$\mathbf{F}_\Sigma; \quad \mathbf{F}_0 = -\mathbf{F}_\Sigma; \quad \mathbf{G}_\Sigma; \quad \mathbf{G}_0 = -\mathbf{G}_\Sigma \tag{7.9}$$

7.2 Structure

Between the terms explained above there are relations which characterize the structure of the general weightless solution consisting of all weightless stress tensor fields with given boundary stresses. This structure is shown in Fig. 7.2 and will be explained in the following.[5]

A-Solution Sets,A-∂_nA-Boundary Fields and T-Solution Sets

A-solution sets to different **A**-∂_n**A**-boundary fields are obviously disjoint. In contrast, the **T**-solution sets to different **A**-∂_n**A**-edge fields or different **A**-solution sets are either disjoint or equal.[6] The construction of the general solution from disjoint **T**-solution sets represents a classification of this general solution, where each **T**-solution set defines a class. This classification can be transferred to the **A**-solution sets and the **A**-∂_n**A**-boundary fields, where a class consists of those **A**-solution sets and **A**-∂_n**A**-boundary fields, respectively, which lead to the same **T**-solution set.

---

[5] So that the text does not become too cumbersome, the terms and their symbols will be used interchangeably in the following discussion.

[6] If two **T**-solution-sets, which result from two **A**-solution-sets resp. from two **A**-∂_n**A**-boundary -fields, have a **T**-field in common, then in each of the two **A**-solution-sets there is a **A**-field, which leads to this **T**-field, so that these two **A**-fields differ by a redundancy-function $\mathbf{A}^\bullet$. Then the two **A**-∂_n**A**-boundary-fields can be converted into each other by addition or subtraction of the **A**-∂_n**A**-edge-field of this redundancy-function $\mathbf{A}^\bullet$, thus both **A**-solution-sets can also be converted into each other by addition or subtraction of this redundancy-function $\mathbf{A}^\bullet$ and therefore the two **T**-solution-sets are equal.

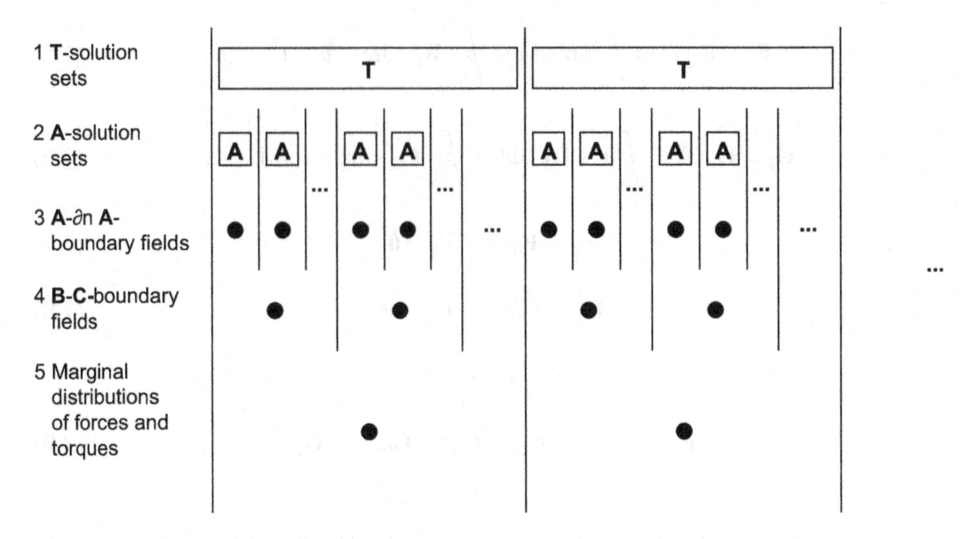

Fig. 7.2 The structure of the general weightless solution of the balance and boundary conditions is represented by the column structure of the figure

The classes of **A**-solution sets and **A**-∂_n**A**-boundary fields can also be defined without considering their **T**-solution sets: Two **A**-solution sets are in a class exactly if they can be merged into each other by addition or subtraction of a redundancy function $\mathbf{A}^{\bullet}$, and two **A**-∂_n**A** boundary fields are in a class exactly if they merge into each other by addition or subtraction of the **A**-∂_n**A** boundary field of a redundancy function $\mathbf{A}^{\bullet}$.

B-C-Boundary Fields

Each **A**-∂_n**A** -boundary field defines a **B**-**C**- boundary field.[7] Since **A**-∂_n**A** boundary fields that define the same **B**-**C** boundary field are in the same class,[8] the classification can also be transferred to the **B**-**C** boundary fields: A class of **B**-**C**-boundary fields is created from all **A**-∂_n**A**-boundary fields of a class.

The classes of **B**-**C**-boundary fields can also be defined without considering the **A**-∂_n**A**-boundary fields: Two **B**-**C**- boundary fields are in a class if and only if they differ from each other by the **B**-**C**- boundary field $\boldsymbol{B}_{\Sigma}^{\bullet}$, $\boldsymbol{C}_{\Sigma}^{\bullet}$ of a redundancy function $\mathbf{A}^{\bullet}$,[9] which is equivalent

[7] The boundary fields $\mathbf{B}_{\Sigma}$ and $\mathbf{C}_{\Sigma}$ are expressed by the boundary values of the **A**-field and its first derivatives. These first derivatives can be expressed by the derivatives of the **A**-field parallel and perpendicular to the boundary surface Σ, that is, by derivatives of the boundary field $\mathbf{A}_{\Sigma}$ parallel to the boundary surface and by the normal derivative $\partial_n\mathbf{A}$. Thus they are defined by the functions $\mathbf{A}_{\Sigma}$and $\partial_n\mathbf{A}$ defined on Σ.

[8] The proof is given in Sect. 15.3.

[9] Justification: Two **B**-**C**-boundary fields lie in a class if and only if the two **A**-∂_n**A**-boundary fields lie in a class from which they arise and these differ by the **A**-∂_n**A**-boundary field of a redundancy

to saying that they differ from each other by a gradient field.[10] Thus, this classification of **B**-**C**- boundary fields defined on the boundary surface Σ can also be described by saying that two **B**-**C**- boundary fields belong to the same class exactly if their orbital integrals[11] coincide over every closed path running on the boundary surface Σ.

Not only the classes of the **B**-**C** – boundary fields, but also these boundary fields themselves can be defined without considering the $\mathbf{A}$-$\partial_n\mathbf{A}$ – boundary fields. The integrals of the **B**-**C**-boundary-fields $\mathbf{B}_\Sigma$ and $\mathbf{C}_\Sigma$ over the oriented edge-curves $\partial\Gamma$ of oriented partial-surfaces Γ of the boundary-surface Σ must namely agree with the forces (7.10) and torques (7.11) respectively, which are generated by the boundary-stresses **t** on these partial-surfaces Γ. According to Stokes' theorem, this can also be expressed by differential equations (7.12) and (7.13).[12] These properties of **B**-**C**-boundary fields are not only a necessary condition, but also a sufficient condition in the following sense: If two boundary fields $\mathbf{B}_\Sigma$, $\mathbf{C}_\Sigma$ satisfy these differential equations, then they form the **B**-**C**-boundary field of $\mathbf{A}$-$\partial_n\mathbf{A}$-boundary fields.[13] Thus the **B**-**C**-boundary fields can be characterized by these differential equations (7.12) and (7.13) alone.

Marginal Distributions of Forces and Torques

Since different **B**-**C**-boundary fields are in the same class if and only if their path integrals over any closed curves on the boundary surface Σ coincide, all **B**-**C**-boundary fields from a class produce the same marginal distribution (7.1)–(7.6) of forces and torques. The converse is also true: if two **B**-**C**-edge fields produce the same marginal distribution of forces and torques, then they are in the same class.[14] So the marginal distributions of forces and torques characterize the classes of the general solution. The **T**-solution set of a class consists of those weightless solutions of the balance and boundary conditions which produce the same marginal distribution of forces and torques, namely the marginal distribution of this class.

function $\mathbf{A}^\bullet$, so that the two **B**-**C**-boundary fields differ by the **B**-**C**-boundary field $\mathbf{B}^\bullet_\Sigma$, $\mathbf{C}^\bullet_\Sigma$ of this redundancy function.

[10] This gradient field consists of the fields $\mathbf{B}^\bullet_\Sigma$ (6.23) and $\mathbf{C}^\bullet_\Sigma$ (6.24), which are gradients of vector fields.

[11] Meant are the orbital integrals of the boundary fields $\mathbf{B}_\Sigma$ and $\mathbf{C}_\Sigma$, of which the respective **B**-**C**-boundary field consists.

[12] In (7.12) and (7.13) only derivatives tangent to the boundary surface Σ occur, defined by the boundary values $\mathbf{B}_\Sigma$ and $\mathbf{C}_\Sigma$ alone.

[13] See Sect. 15.3.

[14] The path integrals over the oriented boundary curves $\partial\Gamma$ of any partial surfaces Γ of the boundary surface Σ coincide by assumption, since they are equal to the forces and torques generated by the boundary stresses on the partial surfaces Γ. Because of the same marginal distribution of forces and torques, the integrals over the boundary curves of the connected surfaces on which no boundary conditions are set must also agree. Thus, the integrals over any closed paths on the boundary surface Σ also agree.

If the boundary surface Σ contains no multiply connected parts, then there is only one class, only one marginal distribution (7.9) of forces and torques and no free parameters.

$$\oint_{\partial\Gamma} \mathbf{B}_\Sigma \cdot d\mathbf{r} = \int_\Gamma \mathbf{t} \cdot dA; \quad \Gamma \overset{\text{prec.}}{\subseteq} \Sigma \tag{7.10}$$

$$\oint_{\partial\Gamma} \mathbf{C}_\Sigma \cdot d\mathbf{r} = \int_\Gamma \mathbf{r} \times \mathbf{t} \cdot dA; \quad \Gamma \overset{\text{prec.}}{\subseteq} \Sigma \tag{7.11}$$

$$(\text{rot}\mathbf{B}_\Sigma)^{\mathrm{T}} \cdot \mathbf{n} \overset{\text{def}}{=} \downarrow \mathbf{B}_\Sigma \cdot (\mathbf{n} \times \nabla) = \mathbf{t} \tag{7.12}$$

$$(\text{rot}\mathbf{C}_\Sigma)^{\mathrm{T}} \cdot \mathbf{n} \overset{\text{def}}{=} \downarrow \mathbf{C}_\Sigma \cdot (\mathbf{n} \times \nabla) = \mathbf{r}_\Sigma \times \mathbf{t} \tag{7.13}$$

7.3 Construction

Construction of all weightless stress tensor fields with given boundary stresses – these stress tensor fields form the general weightless solution of the balance and boundary conditions for weightless stress tensor fields – means to give for each marginal distribution of forces and torques all solutions **T** of the balance and boundary conditions (2.21)–(2.23) for weightless stress tensor fields, which produce this marginal distribution of forces and torques.[15] This marginal distribution is defined by the values of the free parameters, which denote the forces (7.5) and torques (7.6) on all – except one – separate, connected boundary surfaces on which no boundary conditions are imposed. So these solutions consist of all weightless stress tensor fields **T** (6.9), which generate these forces (7.5) and torques (7.6) as well as the given boundary stresses **t** (2.23). If there is no multiple connection[16] on the boundary surface Σ with given boundary stresses, then there are no free parameters and the corresponding relations for the forces (7.5) and torques (7.6) are omitted.

These weightless stress tensor fields **T** and the matrix fields **A** from which they arise by derivation (6.9) are written as the sum of three summands $\mathbf{T}_*, \mathbf{T}_{**}, \mathbf{T}_0$ (7.15) and $\mathbf{A}_{**}, \mathbf{A}_*, \mathbf{A}_0$ (7.14) respectively. Their contributions to the general weightless solution **T** are as follows[17]:

[15] See Fig. 7.2, where the marginal distribution of forces and torques is in row five and all solutions **T** that produce this marginal distribution form the **T**-solution set in row 1.

[16] See Fig. 7.1a, b.

[17] Each of the **A**-fields $\mathbf{A}, \mathbf{A}_{*}, \mathbf{A}_{**}\ \mathbf{A}_0$ defines its corresponding **B** – or **C** – or **T**-field by (6.8) or (6.10) or (6.9), respectively.

- The weightless stress tensor field $\mathbf{T}_*$ is a special solution of the balance and boundary conditions. It generates the given boundary stresses $\mathbf{t}$ (7.16) on the oriented boundary surface Σ and has vanishing free parameters (7.17) and (7.18).
- The weightless stress tensor field $\mathbf{T}_{**}$ generates no boundary stresses (7.19) on the boundary surface Σ and the forces (7.20) and torques (7.21) on the connected boundary surfaces without boundary conditions, which can be specified as free parameters at will.
- $\mathbf{T}_0$ stands for all weightless stress tensor fields which do not produce boundary stresses on the boundary surface Σ (7.22) and whose free parameters vanish (7.23) and (7.24), i.e. which do not influence either the boundary stresses or the free parameters. One obtains these stress tensor fields $\mathbf{T}_0$ as $\mathbf{T}$-fields of $\mathbf{A}$-fields $\mathbf{A}_0$, which together with their first derivatives vanish on the boundary surface Σ (7.25) and are otherwise arbitrary.[18]

This form of the general weightless solution $\mathbf{T}$ can be embedded into the structure of the general solution. The $\mathbf{T}$-solution set generated by all variants of $\mathbf{T}_0$-fields consists of all weightless stress tensor fields $\mathbf{T}$ (7.15) which generate the given boundary stresses $\mathbf{t}$ (2.23) and have the free parameters (7.5) and (7.6) of $\mathbf{T}_{**}$ (7.20) and (7.21).[19]

This gives a simple representation (7.15) of the general weightless solution consisting of all weightless stress tensor fields $\mathbf{T}$ with given boundary stresses $\mathbf{t}$.[20] Their dependence on the free parameters is in the summand $\mathbf{T}_{**}$, which is a linear function of these parameters. The summand $\mathbf{T}_0$ stands for a simply defined class of functions, namely the $\mathbf{T}$-fields (6.9) of $\mathbf{A}$-matrix fields $\mathbf{A}_0$, which together with their first derivatives vanish on the boundary surface Σ (7.25) and are otherwise arbitrary.

—

$$\mathbf{A} = \mathbf{A}_* + \mathbf{A}_{**} + \mathbf{A}_0 \tag{7.14}$$

$$\mathbf{T} = \mathbf{T}_* + \mathbf{T}_{**} + \mathbf{T}_0 \tag{7.15}$$

* * *

[18] That all such stress tensor fields $\mathbf{T}_0$ are obtained this way follows from the structure of the general solution for the case of vanishing boundary stresses $\mathbf{t}$. In this case the vanishing $\mathbf{A}$-$\partial_n\mathbf{A}$-boundary field in row 3 of Fig. 7.2 fits vanishing marginal distribution in row 5 and thus the solution set $\mathbf{T}$ consisting of all fields $\mathbf{T}_0$ in row 1 results from the $\mathbf{A}$-solution set in row 2 consisting of all $\mathbf{A}$-fields which vanish together with their first derivatives on the boundary surface Σ.

[19] This $\mathbf{T}$-solution set is in line 1 of Fig. 7.2. It arises from a $\mathbf{A}$-solution set in line 2. This $\mathbf{A}$-solution set consists of all $\mathbf{A}$-fields (7.14) obtained by all variants of $\mathbf{A}_0$-fields. This $\mathbf{A}$-solution set in line 2 fits the marginal distribution of forces and torques in line 5, which is defined by the free parameters of the summand $\mathbf{A}_{**}$.

[20] The $\mathbf{A}$-fields and $\mathbf{A}_*$ and $\mathbf{A}_{**}$ and their $\mathbf{T}$-fields resp. $\mathbf{T}_*$ and $\mathbf{T}_{**}$ are calculated in Chap. 16.

$$\oint_{\partial\Gamma} \mathbf{B}_* \cdot \mathbf{dr} = \int_{\Gamma} \mathbf{T}_*\mathbf{n} \cdot dA = \int_{\Gamma} \mathbf{t} \cdot dA; \quad \Gamma \text{ prec.} \subseteq \Sigma \tag{7.16}$$

$$\oint_{\partial\Lambda_\nu} \mathbf{B}_* \cdot \mathbf{dr} = \int_{\Lambda_\nu} \mathbf{T}_*\mathbf{n} \cdot dA = \mathbf{0}; \quad \nu = 1, \ldots, \mathrm{n} \tag{7.17}$$

$$\oint_{\partial\Lambda_\nu} \mathbf{C}_* \cdot \mathbf{dr} = \int_{\Lambda_\nu} \mathbf{r} \times \mathbf{T}_*\mathbf{n} \cdot dA = \mathbf{0}; \quad \nu = 1, \ldots, \mathrm{n} \tag{7.18}$$

* * *

$$\oint_{\partial\Gamma} \mathbf{B}_{**} \cdot \mathbf{dr} = \int_{\Gamma} \mathbf{T}_{**}\mathbf{n} \cdot dA = \mathbf{0}; \quad \Gamma \text{ prec.} \subseteq \Sigma \tag{7.19}$$

$$\oint_{\partial\Lambda_\nu} \mathbf{B}_{**} \cdot \mathbf{dr} = \int_{\Lambda_\nu} \mathbf{T}_{**}\mathbf{n} \cdot dA = \mathbf{F}_\nu; \quad \nu = 1, \ldots, \mathrm{n} \tag{7.20}$$

$$\oint_{\partial\Lambda_\nu} \mathbf{C}_{**} \cdot \mathbf{dr} = \int_{\Lambda_\nu} \mathbf{r} \times \mathbf{T}_{**}\mathbf{n} \cdot \mathrm{dA} = \mathbf{G}_\nu; \quad \nu = 1, \ldots, \mathrm{n} \tag{7.21}$$

* * *

$$\oint_{\partial\Gamma} \mathbf{B}_0 \cdot \mathbf{dr} = \int_{\Gamma} \mathbf{T}_0\mathbf{n} \cdot \mathrm{d}A = \mathbf{0}; \quad \Gamma \text{ prec.} \subseteq \Sigma \tag{7.22}$$

$$\oint_{\partial\Lambda_\nu} \mathbf{B}_0 \cdot \mathrm{d}\boldsymbol{r} = \int_{\Lambda_\nu} \mathbf{T}_0\mathbf{n} \cdot \mathrm{d}A = \mathbf{0}; \quad \nu = 1, \ldots, n \tag{7.23}$$

$$\oint_{\partial\Lambda_\nu} \mathbf{C}_0 \cdot \mathbf{dr} = \int_{\Lambda_\nu} \mathbf{r} \times \mathbf{T}_0\mathbf{n} \cdot \mathrm{dA} = \mathbf{0}; \quad \nu = 1, \ldots, \mathrm{n} \tag{7.24}$$

$$\mathbf{A}_{0\Sigma} = \mathbf{0}; \quad \partial_n \mathbf{A}_0 = \mathbf{0} \tag{7.25}$$

7.4 Redundancies and Normalizations

The summand $\mathbf{T}_0$ of the general weightless solution (7.15) stands for the **T**-fields (6.9) of **A**-fields $\mathbf{A}_0$, which together with their first derivatives vanish on the boundary surface Σ of given boundary stresses and are otherwise arbitrary. Redundancies occur among these $\mathbf{A}_0$-fields, since different $\mathbf{A}_0$-fields can lead to the same $\mathbf{T}_0$-field.[21]

These redundancies can be reduced by restricting general solution (7.15) to symmetric fields $\mathbf{A}_0$, since the antisymmetric components have no effect on the fields $\mathbf{T}_0$. However, redundancies still occur among these symmetric fields $\mathbf{A}_0$.[22] One can reduce these redundancies by normalization,[23] but the normalized **A**-fields produced in this process in general no longer satisfy the homogeneous boundary conditions (7.25), so they are no longer $\mathbf{A}_0$-fields. The reduction of redundancies by normalization would therefore generally lead to more complicated boundary conditions, so that normalization is not necessarily advantageous.

However, in the case of symmetrical $\mathbf{A}_0$-fields, a normalization[24] can be carried out in such a way that $\mathbf{A}_0$-fields result again, if the normalization directions of this normalization given in the following are transverse[25] to the boundary surface Σ.[26]

normalization type	normalization	normalization direction
A	$xx - yy - zz$	x,y,z
B	$xx - yy - xy$	z
C	$xx - yy - xz$	y,z
D	$xx - xy - yz$	y,z
E	$xy - yz - xz$	x,y,z

(7.26)

For example, *xx-yy-xz* normalization of all $\mathbf{A}_0$-matrix fields to symmetric $\mathbf{A}_0$-matrix fields with only three independent nonvanishing components $A_{0\ xx}$, $A_{0\ yy}$ and $A_{0\ xz}$ is possible if the boundary surface Σ is transverse to the y- and z-directions, which is equivalent to this boundary surface Σ being representable by both a function $z = z_\Sigma(x, y)$ and a function $y = y_\Sigma(x, z)$.

---

[21] In this case, the different $\mathbf{A}_0$-fields are each distinguished by a redundancy function (6.20) which vanishes together with its first derivatives on the boundary surface Σ.

[22] Redundancies occur when different symmetric $\mathbf{A}_0$-fields differ by a symmetric redundancy function (6.21) which vanishes together with its first derivatives on the boundary surface Σ.

[23] See Sect. 6.2.2.

[24] See Sect. 6.2.2.

[25] A surface and a direction are transverse to each other if each straight line parallel to this direction intersects the surface at most once. For normalization directions, x denotes the direction parallel to the x-axis, and so on.

[26] This is shown in Sect. 14.3.

Both the orientation of the coordinate system and the shape of the boundary surface Σ have influence on whether the conditions for a normalization of the $\mathbf{A}_0$-fields are fulfilled. If these conditions are fulfilled, then the general solution (7.15) can be designed very simply, by representing all stress tensor fields $\mathbf{T}_0$ as $\mathbf{T}$-fields (6.9) of $\mathbf{A}_0$-fields, which show this normalization. Thus all stress tensor fields $\mathbf{T}_0$ can be expressed by three scalar functions which together with their first derivatives vanish on the boundary surface Σ, namely the three independent matrix components of the normalized $\mathbf{A}_0$-fields.

These normalization requirements do not usually apply to floating glaciers, since for these glaciers the boundary surface Σ of given boundary stresses consists of the free surface and the underside lying in the water, so that there is usually no direction that is transverse to the boundary surface Σ. In contrast, for land glaciers the boundary surface of known boundary stresses consists only of the free surface, and usually the coordinate system can be oriented to allow one of the above normalizations.

The General Solution of the Balance and Boundary Conditions

8

Abstract

The general solution of the balance and boundary conditions can be constructed as the sum of a special solution of the balance conditions and the general "weightless solution" of the balance and boundary conditions for "weightless stress tensor fields". Thus, the general solution is known, since both special solutions of the balance conditions and the general "weightless solution" are known. If the boundary surface of known boundary stresses consists only of the simply connected free surface, which is usually the case for land glaciers, then the general solution can be obtained from three selected components of the stress tensor or the deviatoric stress tensor, which can be taken as arbitrary functions in the framework of the general solution.

8.1 Representations with Stress Functions

The representations of the general solution by stress functions are characterized by the fact that weightless stress tensor fields, which are components of this general solution, are given as symmetrized twice rotations of matrix fields.[1] Gurtin [2, p. 54] calls such matrix fields "stress functions".

The general solution **S** of the balance and boundary conditions (2.14)–(2.16) can be constructed as a sum (8.1) of a special solution of the balance conditions $\mathbf{S}_{\text{bal}}$ (2.18) and (2.19) and of the general weightless solution **T** (7.15) of the balance and boundary conditions for weightless stress tensor fields (2.21)–(2.23). Thus the general solution **S** is

[1] According to (6.9), this is the general form of weightless stress tensor fields.

P. Halfar, *Stresses in glaciers*, https://doi.org/10.1007/978-3-662-66024-9_8

known, since both stress tensor fields $\mathbf{S}_{\text{bal}}$[2] and the general weightless solution $\mathbf{T}$ (7.15) are known.

If multiple connection[3] occurs on the boundary surface Σ of given boundary stresses, then there are free parameters in the general weightless solution $\mathbf{T}$ and thus also in the general solution $\mathbf{S}$, whose values can be chosen arbitrarily. These free parameters are the forces $\mathbf{F}_\nu[\mathbf{T}]$ (7.5) and torques $\mathbf{G}_\nu[\mathbf{T}]$ (7.6) generated by the general weightless solution $\mathbf{T}$ on the separate, connected surfaces Λ_ν – except one of them –, where no boundary conditions occur. Alternatively, one can also use the forces $\mathbf{F}_\nu[\mathbf{S}]$ and torques $\mathbf{G}_\nu[\mathbf{S}]$ (8.2) generated by the general solution $\mathbf{S}$ as free parameters, which are obtained from the forces and torques generated by the weightless stress tensor field $\mathbf{T}$ by adding (8.4) the forces $\mathbf{F}_\nu[\mathbf{S}_{\text{bal}}]$ and torques $\mathbf{G}_\nu[\mathbf{S}_{\text{bal}}]$ (8.3) generated by the stress tensor field $\mathbf{S}_{\text{bal}}$. When determining a realistic stress tensor field $\mathbf{S}$, these force parameters and torque parameters $\mathbf{F}_\nu[\mathbf{S}]$ and $\mathbf{G}_\nu[\mathbf{S}]$ must be specified in such a way that they correspond to the corresponding real forces and torques.

The general solution $\mathbf{S}$ (8.1) consists of the four summands $\mathbf{S}_{\text{bal}}$, $\mathbf{T}_*$, $\mathbf{T}_{**}$ and $\mathbf{T}_0$, which can be calculated by integrations and differentiations. These summands have the following properties:

1. $\mathbf{S}_{\text{bal}}$[4] is a solution of the balance conditions (2.18) and (2.19).
2. $\mathbf{T}_*$ results from a stress function $\mathbf{A}_*$.[5] This weightless stress tensor field $\mathbf{T}_*$ generates the given boundary stresses $\mathbf{t}$ (2.24) and (7.16) on Σ and has vanishing free parameters (7.17) and (7.18).

$\mathbf{T}_*$ is a special solution $\mathbf{T}_{\text{spez}}$ of the balance and boundary conditions (2.21)–(2.23) for weightless stress tensor fields, which does not generate forces and torques (7.17) and (7.18) on all – except one – connected boundary surfaces Λ_ν, on which no boundary conditions are set. Consequently, the sum of $\mathbf{S}_{\text{bal}}$ and $\mathbf{T}_*$ is a special solution $\mathbf{S}_{\text{spez}}$ of the balance and boundary conditions (2.14)–(2.16), which generates the same forces and torques (8.3) on the boundary surfaces Λ_ν as $\mathbf{S}_{\text{bal}}$.

[2] For example, $\mathbf{S}_b$ (5.4) or $\mathbf{S}_e$ (5.13) are such stress tensor fields.

[3] S. Fig. 7.1 c, d.

[4] As a solution $\mathbf{S}_{\text{bal}}$ of the balance conditions (2.18) and (2.19) one can use for example one of the stress tensor fields $\mathbf{S}_b$ (5.4) or $\mathbf{S}_e$ (5.13).

[5] The calculation of $\mathbf{A}_*$ and $\mathbf{T}_*$ is described in Sect. 16.1. The symmetrized twofold rotations of matrix fields occurring in (8.1) are identical to the twofold rotations of the symmetrized matrix fields according to (13.18).

3. $\mathbf{T}_{**}$ results from a stress function $\mathbf{A}_{**}$, which is a linear function of the free parameters $\mathbf{F}_\nu[\mathbf{T}]$ and $\mathbf{G}_\nu[\mathbf{T}]$.[6] This stress tensor field $\mathbf{T}_{**}$ is a variable part of the general solution, which is weightless, generates no boundary stresses on the boundary surface Σ (7.19), and generates the forces $\mathbf{F}_\nu[\mathbf{T}]$ (7.20) and torques $\mathbf{G}_\nu[\mathbf{T}]$ (7.21) on the connected boundary surfaces Λ_ν, which can be given as free parameters at will.
4. $\mathbf{T}_0$ arises from a stress function $\mathbf{A}_0$,[7] which together with its first derivatives vanishes on the boundary surface Σ of given boundary stresses and is otherwise arbitrary. This stress tensor field $\mathbf{T}_0$ is a variable part of the general solution, which is weightless, has vanishing boundary stresses on the surface Σ (7.22) and has vanishing free parameters (7.23) and (7.24). Therefore $\mathbf{T}_0$ generates no forces and torques on all connected boundary surfaces Λ_ν on which no boundary conditions are set.

One can restrict $\mathbf{A}_0$ to normalized matrix fields, which are symmetrical and for which according to normalization three of their six independent matrix elements are zero, if the shape of the boundary surface Σ and the orientation of the coordinate system fulfill corresponding conditions.[8]

So all stress tensor fields $\mathbf{S}$ (8.1) of the general solution are obtained by variations of their variable parts $\mathbf{T}_{**}$ and $\mathbf{T}_0$ by varying the free parameters $\mathbf{F}_\nu[\mathbf{T}]$ and $\mathbf{G}_\nu[\mathbf{T}]$ arbitrarily and by varying the stress function $\mathbf{A}_0$ of the summand $\mathbf{T}_0$ arbitrarily.[9] A realistic solution is obtained by specifying realistic values for the forces $\mathbf{F}_\nu[\mathbf{S}]$ and torques $\mathbf{G}_\nu[\mathbf{S}]$ (8.2) and thereby also defining the values of the free parameters $\mathbf{F}_\nu[\mathbf{T}]$ and $\mathbf{G}_\nu[\mathbf{T}]$ (8.4) and by selecting the matrix function $\mathbf{A}_0$ appropriately.[10]

---

[6] The calculation of $\mathbf{A}_{**}$ and $\mathbf{T}_{**}$ is made in Sect. 16.2.

[7] One can restrict $\mathbf{A}_0$ to symmetric matrix fields in (8.1), since their antisymmetric components do not contribute to $\mathbf{T}_0$.

[8] The possible normalizations and their prerequisites are set out in Sect. 7.4.

[9] The arbitrary variations of matrix fields $\mathbf{A}_0$ are by definition subject to the restriction that all matrix fields $\mathbf{A}_0$ together with their first derivatives vanish on the boundary surface Σ of given boundary stresses. One can restrict oneself to normalized variations if the corresponding normalization conditions are fulfilled.

[10] As mentioned above, the selection of a realistic solution is no longer the subject of this general investigation. Such a selection can only be made on a case-by-case basis and requires specific data collected in the field.

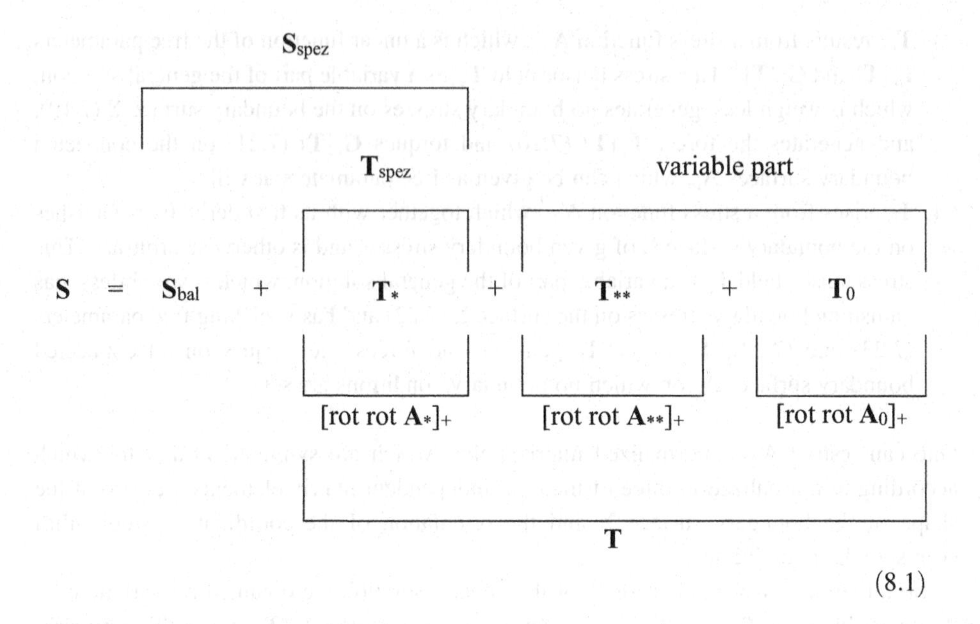

(8.1)

* * *

$$\nu = 1, \ldots, n :$$

$$\mathbf{F}_\nu[\mathbf{S}] \stackrel{\text{def}}{=} \int_{\Lambda_\nu} \mathbf{S}\mathbf{n} \cdot dA; \quad \mathbf{G}_\nu[\mathbf{S}] \stackrel{\text{def}}{=} \int_{\Lambda_\nu} \mathbf{r} \times \mathbf{S}\mathbf{n} \cdot dA \tag{8.2}$$

$$\mathbf{F}_\nu[\mathbf{S}_{\text{bal}}] \stackrel{\text{def}}{=} \int_{\Lambda_\nu} \mathbf{S}_{\text{bal}}\mathbf{n} \cdot dA; \quad \mathbf{G}_\nu[\mathbf{S}_{\text{bal}}] \stackrel{\text{def}}{=} \int_{\Lambda_\nu} \mathbf{r} \times \mathbf{S}_{\text{bal}}\mathbf{n} \cdot dA \tag{8.3}$$

$$\mathbf{F}_\nu[\mathbf{S}] = \mathbf{F}_\nu[\mathbf{S}_{\text{bal}}] + \mathbf{F}_\nu[\mathbf{T}]; \qquad \mathbf{G}_\nu[\mathbf{S}] = \mathbf{G}_\nu[\mathbf{S}_{\text{bal}}] + \mathbf{G}_\nu[\mathbf{T}] \tag{8.4}$$

8.2 Representations with Three Independent Stress Components

8.2.1 Problem Definition and Solution Procedure

In this section it is assumed that the boundary surface Σ of given boundary stresses consists of the simply connected free surface of the considered glacier area, where the boundary stresses disappear, and consequently also the variable component $\mathbf{T}_{**}$ of the general solution $\mathbf{S}$ (8.1) disappears. Moreover, normalizability is assumed, so the free surface Σ

is supposed to be transverse to the normalization directions[11] of a normalization, so that the variable component $\mathbf{T}_0$ in this general solution $\mathbf{S}$ arises from arbitrary variations of corresponding normalized $\mathbf{A}_0$-fields.[12]

These arbitrary variations of the normalized $\mathbf{A}_0$-fields are equivalent to arbitrary variations of their three independent matrix elements. Since these three matrix elements have relatively abstract meaning, three scalar functions with concrete meaning shall be used alternatively, namely three selected components of the stress tensor $\mathbf{S}$ or the deviatoric stress tensor $\mathbf{S}'$. By definition, these three selected components are called independent stress components if, given any of these three scalar functions, there is exactly one stress tensor $\mathbf{S}$ with these independent stress components that satisfies the balance and boundary conditions.

The problem of expressing the stress tensor field $\mathbf{S}$ by its three selected independent components can be reduced to the problem of expressing the weightless stress tensor field $\mathbf{T}_0$[13] by its three selected independent stress components as follows: Compute a special solution $\mathbf{S}_*$, where the three selected stress components vanish,[14] add to it the weightless stress tensor field $\mathbf{T}_0$, and thus have the solution $\mathbf{S}$ (8.5) of the original problem, if the independent stress components of $\mathbf{T}_0$ are chosen to be the independent stress components of $\mathbf{S}$.[15]

To express the weightless stress tensor field $\mathbf{T}_0$ by its three selected independent stress components, one represents $\mathbf{T}_0$ by a normalized $\mathbf{A}_0$-field defined by three independent $\mathbf{A}_0$-matrix elements which vanish together with their first derivatives at the free surface Σ (8.7). The three independent stress components of $\mathbf{T}_0$ are then also expressed by the three independent $\mathbf{A}_0$-matrix elements. If one reverses these relations, then one has the entire normalized matrix field $\mathbf{A}_0$ and thus also the matrix field $\mathbf{T}_0$ (8.7) expressed by the three independent stress components of $\mathbf{T}_0$.

To calculate the three independent $\mathbf{A}_0$-matrix elements a system of three partial differential equations has to be solved with the boundary condition that these three $\mathbf{A}_0$-matrix elements and their first derivatives vanish at the free surface Σ. This problem can be formulated as a matrix differential equation with boundary conditions (3.32) and (3.33) by writing the three sought matrix elements of $\mathbf{A}_0$ as components of a three-row column matrix function $\mathbf{f}$, introducing a three-row column matrix function $\mathbf{q}$ whose components are

[11] See Sect. 7.4.

[12] See point 4, Sect. 8.1.

[13] The index "0" means that $\mathbf{T}_0$ can be obtained from a stress function $\mathbf{A}_0$, which together with its first derivatives disappears at the free surface Σ. Thus, the boundary stresses of $\mathbf{T}_0$ disappear at the free surface Σ. According to the precondition $\mathbf{A}_0$ is normalized.

[14] To do this, one starts from some special solution $\mathbf{S}_{spez}$ and subtracts a weightless solution $\mathbf{T}_0$, for whose independent stress components one substitutes the corresponding components of $\mathbf{S}_{spez}$.

[15] This is also true for independent deviatoric components because of relation (8.6).

defined by the independent stress components of $\mathbf{T}_0$, and using a quadratic matrix operator $\mathcal{L}$ whose nine matrix elements each consist of second-order differential operators.[16]

$$\mathbf{S} = \mathbf{S}_* + \mathbf{T}_0 \tag{8.5}$$

$$\mathbf{S}' = \mathbf{S}'_* + \mathbf{T}'_0 \tag{8.6}$$

$$\mathbf{T}_0 = \text{rot rot } \mathbf{A}_0; \quad \mathbf{A}_0 \text{ normalized}; \quad \mathbf{A}_{0\Sigma} = \partial_n \mathbf{A}_0 = \mathbf{0} \tag{8.7}$$

8.2.2 Calculation of the Solutions

The special solution $\mathbf{S}_*$, the weightless stress tensor field $\mathbf{T}_0$ and hence the representation of the general solution $\mathbf{S}$ (8.5) by three independent stress components are computed for eight different combinations[17] "a" to "h" of three independent stress components each.[18] The following convex cones play a role:

- K_x: One-dimensional cone or ray generated by the base and cone vector $\mathbf{e}_x$. K_y and K_z are defined analogously.
- K_{xy}: Two-dimensional cone generated by the two base and cone edge vectors $\mathbf{e}_x$ and $\mathbf{e}_y$. K_{yz} and K_{xz} are defined analogously.
- K_{xyz}: Three-dimensional cone generated by the three base and cone edge vectors $\mathbf{e}_x$, $\mathbf{e}_y$ and $\mathbf{e}_z$.
- $K_z^{\odot}$: Three-dimensional rotationally symmetric cone with the opening angle $\pi/2$, whose rotation semi-axis points from the cone apex in the positive z-direction. $K_y^{\odot}$ is defined analogously.
- K'_{xz}: Two-dimensional cone generated by its two edge vectors $\mathbf{e}_x + \sqrt{2}\mathbf{e}_z$ and $-\mathbf{e}_x + \sqrt{2}\mathbf{e}_z$.
- K''_{xz}: Two-dimensional cone generated by its two edge vectors $\mathbf{e}_x + \sqrt{2}\mathbf{e}_z$ and $\mathbf{e}_x - \sqrt{2}\mathbf{e}_z$. The two cones K'_{xz} and K''_{xz} complement each other to form a half-plane.

[16] Such matrix differential equations and methods for their solution are discussed in Sect. 3.4. The differential equations appearing in this paper and their solutions are given in Chap. 17.

[17] An analysis of all possible combinations of three independent stress components is carried out in Chap. 17.

[18] $\mathbf{S}_*$ and $\mathbf{T}_0$ are given in Chap. 17 for each of the combinations "a" to "h".

- K'_{yxz}: Three-dimensional cone generated by its three edge vectors $\mathbf{e}_y$, $\mathbf{e}_x + \sqrt{2}\mathbf{e}_z$ and $-\mathbf{e}_x + \sqrt{2}\mathbf{e}_z$.
- K''_{yxz}: Three-dimensional cone generated by its three edge vectors $\mathbf{e}_y$, $\mathbf{e}_x + \sqrt{2}\mathbf{e}_z$ and $\mathbf{e}_x - \sqrt{2}\mathbf{e}_z$.

The model "h" with three independent deviatoric *xy*-, *yz*- and *xx*-components is a special case, because in this model two mutually exclusive representations of the general solution **S** with different model and integration cones are possible.[19] Table 8.1 gives an overview of the eight models "a" to "h".

Thus, simply structured representations of the general solution **S** (8.5) of the balance and boundary conditions with different combinations "a" to "h" of three independent stress components are available. These representations are created by applying differential and integral operators to given functions. The component $\mathbf{S}_*$ of this general solution **S** (8.5) is obtained by applying these operators to the ice density and the component $\mathbf{T}_0$ is obtained by applying these operators to the three independent stress components.[20]

With these operator representations[21] of the general solution, the balance and boundary conditions (2.14)–(2.16) can be checked as follows: The balance relations (2.14) and (2.15) follow from the calculus rules for the operators.[22] That the boundary stresses of $\mathbf{T}_0$ vanish at the free surface Σ follows from the fact that the stress function $\mathbf{A}_0$ and its first derivatives vanish there, since in the formulas for $\mathbf{A}_0$ all operator products contain a corresponding excess number of integral operators.[23] That not only the boundary stresses of $\mathbf{S}_*$ but also $\mathbf{S}_*$ itself disappear at the free surface Σ also follows from a corresponding excess number of integral operators in all operator products.

8.2.3 Surface Shape and Domain of Definition

As a prerequisite for the existence of solutions $\mathbf{T}_0$ and $\mathbf{S}_*$ the free surface Σ must be transverse and synchronous to the respective model cone.[24] This precondition does not only concern the shape of the free surface Σ, but also the orientation of the coordinate system, since a rotation of the coordinate system causes a corresponding rotation of the

[19] S. Chap. 17.

[20] The formula numbers for $\mathbf{S}_*$ and $\mathbf{T}_0$ are given in Table 8.1.

[21] See the formulae for $\mathbf{A}_0$, for $\mathbf{T}_0$ and for $\mathbf{S}_* = \mathbf{S}_a$ to $\mathbf{S}_* = \mathbf{S}_h$ in Sects. 17.1, 17.2, 17.3, 17.4, 17.5, 17.6, 17.7, and 17.8.

[22] See Sect. 3.2, formulae (3.21)–(3.27).

[23] See Sect. 3.4.3.

[24] See the definition of "transverse and synchronous" in Sect. 3.4.1. The model cones are given for each of the models "a" to "h" both in Table 8.1 and in Sects. 17.1, 17.2, 17.3, 17.4, 17.5, 17.6, 17.7, and 17.8.

Table 8.1 Models with the combinations "a" to "h" of three selected independent stress components

	a	b	c	d	e	f	g	h	
Model-Cone	K_{xyz}	K_z	K_{yz}	K_{yz}	K_{xyz}	$K^{\odot}_{z}$	$K^{\odot}_{y}$	K'_{yxz}	K''_{yxz}
$\mathbf{T_0}$	(17.8)	(17.16)	(17.24)	(17.32)	(17.40)	(17.48)	(17.58)	(17.70)	
$T_{0\ xx}$	•	•	•	•	K_x	$K^{\odot}_{z}$	$K^{\odot}_{y}$	K'_{yxz}	K''_{yxz}
					–	$K^{\odot}_{z}$	$K^{\odot}_{y}$	K'_{yxz}	K''_{yxz}
					K_x	$K^{\odot}_{z}$	$K^{\odot}_{y}$	K'_{xz}	K''_{xz}
$T_{0\ yy}$	•	•	•	–	K_y	$K^{\odot}_{z}$	$K^{\odot}_{y}$	K_y	
				K_y	K_y	$K^{\odot}_{z}$	$K^{\odot}_{y}$	K_y	
				K_y	–	$K^{\odot}_{z}$	$K^{\odot}_{y}$	–	
$T_{0\ zz}$	•	K_z	K_z	K_z	–	$K^{\odot}_{z}$	$K^{\odot}_{y}$	K'_{yxz}	K''_{yxz}
		K_z	K_z	K_z	K_z	$K^{\odot}_{z}$	$K^{\odot}_{y}$	K'_{yxz}	K''_{yxz}
		K_z	K_z	K_z	K_z	$K^{\odot}_{z}$	$K^{\odot}_{y}$	K'_{xz}	K''_{xz}
$T_{0\ xy}$	K_y		K_y	•	•	•	$K^{\odot}_{y}$	•	
	K_x	•	–				$K^{\odot}_{y}$		
	K_{xy}		K_y				$K^{\odot}_{y}$		
$T_{0\ yz}$	K_{yz}	–	K_{yz}	•	•	$K^{\odot}_{z}$	$K^{\odot}_{y}$	•	
	K_z	K_z	K_z			$K^{\odot}_{z}$	$K^{\odot}_{y}$		
	K_y	K_z	K_y			$K^{\odot}_{z}$	$K^{\odot}_{y}$		
$T_{0\ xz}$	K_z	K_z		K_z	•	$K^{\odot}_{z}$	•	K'_{yxz}	K''_{yxz}
	K_{xz}	–	•	K_z		$K^{\odot}_{z}$		K'_{yxz}	K''_{yxz}
	K_x	K_z		–		$K^{\odot}_{z}$		K'_{xz}	K''_{xz}
$T'_{0\ xx}$						•	•	•	
$T'_{0\ yy}$						•	•		
$\mathbf{S_*}$	(17.9)	(17.17)	(17.25)	(17.33)	(17.41)	(17.49)	(17.59)	(17.71)	
$S_{*\ xx}$	0	0	0	0	K_x	$K^{\odot}_{z}$	$K^{\odot}_{y}$	K'_{yxz}	K''_{yxz}
$S_{*\ yy}$	0	0	0	K_y	K_y	$K^{\odot}_{z}$	$K^{\odot}_{y}$	K_y	
$S_{*\ zz}$	0	K_z	K_z	K_z	K_z	$K^{\odot}_{z}$	$K^{\odot}_{y}$	K'_{yxz}	K''_{yxz}
$S_{*\ xy}$	K_{xy}	0	K_y	0	0	0	$K^{\odot}_{y}$	0	
$S_{*\ yz}$	K_{yz}	K_z	K_{yz}	0	0	$K^{\odot}_{z}$	$K^{\odot}_{y}$	0	
$S_{*\ xz}$	K_{xz}	K_z	0	K_z	0	$K^{\odot}_{z}$	0	K'_{yxz}	K''_{yxz}
$S'_{*\ xx}$						0	0	0	
$S'_{*\ yy}$						0	0		

The table is explained by the example "d" of the three independent components "*xx*", "*xy*" and "*yz*". The independent components of the weightless stress tensor field $\mathbf{T_0}$ are indicated by three dots. Its dependent component T_{0yy} is not affected by T_{0xx} but depends only on T_{0xy} and T_{0yz} in the dependence cone K_y. The data for the other two dependent components T_{0zz} and T_{0xz} should be interpreted accordingly. The special stress tensor field $\mathbf{S}_* = \mathbf{S}_d$ has vanishing selected components S_{*xx}, S_{*xy} and S_{*yz} and the other three components S_{*yy}, S_{*zz} and S_{*xz}, depend on the ice density ρ in the dependence cone K_y, K_z and K_z, respectively. All dependence cones generate the convex model cone K_{yz}. The weightless stress tensor field $\mathbf{T}_0$ is expressed by its independent stress components in formula (17.32) and the stress tensor field $\mathbf{S}_* = \mathbf{S}_d$ is given in formula (17.33)

model cone. If necessary, the coordinate system and with it the model cone are to be rotated in such a way that the above-mentioned condition is fulfilled. It shall be taken into account that such a rotation changes the meaning of the three independent stress components.

In the models "h" with three independent deviatoric xy-, yz- and xx-components there are two alternatives with the model cones K'_{yxz} and K''_{yxz}. In these models "h" the coordinate system must be rotated so that the free surface Σ is transverse and synchronous either to the model cone K'_{yxz} or to the model cone K''_{yxz}.[25]

The spatial domain of definition Ω_{def} of these solutions is determined by the model cone. This domain of definition Ω_{def} consists of the glacier region Ω compatible[26] with the free surface Σ and the model cone, also called the internal region Ω_{int}, and of an external region Ω_{ext} beyond the free surface Σ, generated by all model cones with tip on the free surface Σ. On this external region Ω_{ext}, by definition, the independent stress components of $\mathbf{T}_0$ and the ice density ρ vanish. Therefore, this external region does not contain any information, but is only needed to define the integral operators.[27]

8.2.4 Solution Dependency Cones

The solutions $\mathbf{T}_0$ and $\mathbf{S}_*$ are characterized, among other things, by the convex dependence cones of their matrix elements.[28] The matrix elements of $\mathbf{T}_0$ depend on the three independent stress components and have a convex dependence cone with respect to each of the three independent stress components. When calculating a matrix element of $\mathbf{T}_0$ at a point, contributions from the respective independent stress component come only from the

[25] These two alternatives are mutually exclusive. The free surface Σ cannot be transverse and synchronous to both the model cone K'_{yxz} and the model cone K''_{yxz}, since K'_{yxz} contains the cone vector $-\mathbf{e}_x + \sqrt{2}\mathbf{e}_z$ and K''_{yxz} the cone vector $\mathbf{e}_x - \sqrt{2}\mathbf{e}_z$ opposite to it. S. Chap. 17.

[26] See Sect. 3.4.1.

[27] See point 2, Sect. 3.1.

[28] In the calculations of the matrix elements at a point, integral operators or products of integral operators occur, which define the respective dependency cone. See Sect. 3.3.

respective dependence cone with peak at that point.[29] The same is true for the matrix elements of $\mathbf{S}_*$, where these matrix elements depend only on the ice density ρ.[30]

[29]Not only the values of the independent stress components but also their derivatives provide contributions, so it generally depends on the entire function course of the independent stress components in the dependence cone.

[30]The dependence cones for the matrix elements of $\mathbf{T}_0$ and $\mathbf{S}_*$ are given in Table 8.1.

9 Models and Model Selection

Abstract

So far, several models have been presented for the general solution of the balance and boundary conditions. This chapter contains a characterization of these models and criteria for model selection.

9.1 Characterisation of the Models

In all models, the general solution of the balance and boundary conditions (2.14)–(2.16) is a sum of a special solution and a variable component, which stands for the totality of those weightless stress tensor fields with which all individual solutions of the general solution can be generated. This variable component can be expressed either by stress functions or by independent stress components.

9.1.1 Models with Stress Functions

In the models with stress functions,[1] the variable component of the general solution $\mathbf{S}$ (8.1) consists of two variable summands $\mathbf{T}_{**}$ and $\mathbf{T}_0$, which are expressed by stress functions $\mathbf{A}_{**}$ and $\mathbf{A}_0$, respectively. These models can be characterized as follows:

[1] See Sect. 8.1.

P. Halfar, *Stresses in glaciers*, https://doi.org/10.1007/978-3-662-66024-9_9

1. Variable summand $\mathbf{T}_{**}$

   In case of multiple connection of the boundary surface Σ of given boundary stresses, free parameters occur in the general solution. These parameters are the forces and torques acting on the separate connected boundary surfaces – except one of them – where no boundary conditions are imposed. The stress function $\mathbf{A}_{**}$ and thus the variable summand $\mathbf{T}_{**}$ of the general solution depends on these free parameters.[2] The values of these parameters can vary arbitrarily within the framework of the general solution and thus generate the variants $\mathbf{T}_{**}$.

   If a realistic stress tensor field is to be selected, these parameters are to be assigned their realistic values. If no multiple connection occurs on the boundary surface Σ, the variable summand $\mathbf{T}_{**}$ does not occur.
2. Variable summand $\mathbf{T}_0$

   The variants $\mathbf{T}_0$ arise from arbitrary variants of stress functions $\mathbf{A}_0$.[3] Contributions to the stress tensor come only from the symmetric part of a stress function, which is why only symmetric variants $\mathbf{A}_0$ of stress functions must be considered.
3. Redundancies

   Different variants $\mathbf{A}_0$ of stress functions can lead to the same stress tensor field $\mathbf{T}_0$. In this case, there are redundant variants of stress functions $\mathbf{A}_0$.
4. Normalized models

   A normalized model can be used to calculate the general solution if the normalization directions of the normalization[4] used are transverse to the boundary surface Σ of given boundary stresses.

   In a normalized model, only normalized variants $\mathbf{A}_0$ of stress functions are needed.[5] Therefore, the redundancies are less than in a non-normalized model or have even disappeared. These normalized variants $\mathbf{A}_0$ are symmetric and three of their six independent components vanish. Therefore, all variants $\mathbf{T}_0$ can be expressed by three scalar functions which vanish together with their first derivatives on the boundary surface Σ and are otherwise arbitrary, namely the three non vanishing independent components of $\mathbf{A}_0$.
5. Influence of the coordinate system

   When the Cartesian coordinate system is rotated, the meanings of the $\mathbf{A}_0$- and $\mathbf{T}_0$-matrix elements change and the normalization directions rotate also. The selected coordinate system is therefore also a characteristic of the respective model.

---

[2] See No. 3, Sect. 8.1.

[3] By definition, these arbitrary variants $\mathbf{A}_0$ are subject to the constraint that they vanish together with their first derivatives on the boundary surface Σ of given boundary stresses. Therefore the variants $\mathbf{T}_0$ have vanishing boundary stresses $\mathbf{T}_0\,\mathbf{n} = 0$ on Σ.

[4] These normalization directions are given in tabular form in Sect. 7.4.

[5] See Sect. 7.4.

6. Scope of application of the models

   For any shape of the considered glacier area Ω and any shape of the boundary surface Σ of given boundary stresses there is a suitable model. In any case, models without normalization are applicable. Normalized models are applicable if the above mentioned normalization requirements are fulfilled.
7. Calculation of the general solution

   The general solution of the balance and boundary conditions is calculated by integration and differentiation.[6]

9.1.2 Models with Three Selected Independent Stress Components

In the models with three selected independent stress components,[7] the general solution **S** (8.5) of the balance and boundary conditions (2.14)–(2.16) consists of a special solution $\mathbf{S}_*$ and a variable component $\mathbf{T}_0$. This variable component $\mathbf{T}_0$ is expressed by the three selected independent stress components. The models can be characterized as follows:

1. Scope of application of the models

   A model with three selected independent stress components can be applied under the following conditions:
   (a) The oriented boundary surface Σ of given boundary stresses is simply connected and a free surface.[8]
   (b) The oriented boundary surface Σ is transverse and synchronous[9] to the model cone defined by the three selected stress components.[10]
   (c) The glacier area Ω considered is compatible with the boundary surface Σ and the model cone.[11]

[6] Procedures for calculating the four summands that make up the general solution **S** (8.1) are mentioned in Sect. 8.1 under nos. 1–4.

[7] See Sect. 8.2.

[8] Theoretically, such a model is also applicable if the boundary surface Σ of given boundary stresses is not a free surface. In this case, the special solution $\mathbf{S}_*$ is different from the one given in Chap. 17, but the variable component $\mathbf{T}_0$ remains the same. However, this theoretical case has little practical significance, since the following condition 1b is practically satisfied only when the boundary surface Σ is a free surface. In the other practical cases, in fact, the boundary surface Σ of given boundary stresses consists of a free upper surface and a submerged lower surface, which is why the following condition 1b is not fulfilled.

[9] See Sect. 3.4.1.

[10] The model cones are given in Table 8.1, Sect. 8.2.2 as well as in Sects. 17.1, 17.2, 17.3, 17.4, 17.5, 17.6, 17.7, and 17.8.

[11] See Sect. 3.4.1.

2. Special solution $\mathbf{S}_*$ and variable summand $\mathbf{T}_0$[12]
   The three selected components of the special solution $\mathbf{S}_*$ vanish. Therefore, the three selected components of the variants$\mathbf{T}_0$ coincide with the three selected stress components of the general solution **S** (8.5).
   The variants $\mathbf{T}_0$ are expressed by their three selected stress components. These variants $\mathbf{T}_0$ are generated by arbitrary variants of these three stress components and consist of all weightless stress tensor fields with boundary stresses vanishing at the free surface Σ.
3. Mother models
   Each model with three selected independent stress components arises from a mother model with stress functions, in which the variable component $\mathbf{T}_0$ (8.7) of the general solution **S** (8.5) is expressed by normalized[13] stress functions $\mathbf{A}_0$. In this mother model the variants $\mathbf{T}_0$ arise from arbitrary variants $\mathbf{A}_0$.[14]
4. No redundancies
   If the three selected, independent stress components in the considered glacier area Ω are given arbitrarily, then there is in Ω exactly one solution **S** (8.5) of the balance and boundary conditions with these components. In the mother model, there is exactly one normalized $\mathbf{A}_0$-matrix field, which can be expressed by the three selected stress components.[15]
5. Influence of the coordinate system
   Since the model cone rotates with the coordinate system and the significance of the independent stress components also changes, the models are also characterized by the choice of coordinate system.
6. Glacial area under consideration and domain of definition
   The domain of definition Ω_{def} is larger than the considered glacier area Ω and contains a so-called external area Ω_{ext} beyond the free surface Σ, which is created by parallel displacement of the model cone with its tip along the free surface Σ. This external region is only used to define the integral operators, which can be used to make the calculations clear. This external area does not contain any information, because all allowed functions and distributions disappear in this area. All information is therefore contained in the glacier area Ω under consideration, which is also referred to as the internal area.

[12] For the eight combinations "a" to "h" of selected independent stress components, the formula numbers for $\mathbf{S}_*$ and $\mathbf{T}_0$ are given in Table 8.1, Sect. 8.2.2. The formulae are given in Sects. 17.1, 17.2, 17.3, 17.4, 17.5, 17.6, 17.7, and 17.8.

[13] The normalization of the stress function $\mathbf{A}_0$ in the mother model and this stress function itself are given in Chap. 17 for each selection "a" to "h" of independent stress components.

[14] The arbitrary variants $\mathbf{A}_0$ are by definition under the constraint that they vanish together with their first derivatives on the boundary surface Σ.

[15] See footnote 13.

7. No boundary conditions for the independent stress components

   The independent stress components do not have to satisfy boundary conditions. On the other hand, in the mother model with stress functions, the $\mathbf{A}_0$-matrix fields together with their first derivatives must vanish on the boundary surface Σ of given boundary stresses. The latter results by itself if the $\mathbf{A}_0$-matrix fields are calculated with the help of the independent stress components.
8. Calculation of the general solution

   The calculation of both the constituents $\mathbf{S}_*$ and $\mathbf{T}_0$ of the general solution $\mathbf{S}$ (8.5) and of the normalized $\mathbf{A}_0$-matrix field in the mother model is done by differentiations and integrations of the ice density and the three independent stress components.[16]
9. Meaning

   The calculation of the general solution $\mathbf{S}$ from three selected independent stress components represents a formal mathematical procedure. Theoretically, it means that there is a unique solution $\mathbf{S}$ (8.5) of the balance and boundary conditions for any given independent stress components. Practically, it means that a realistic stress tensor field is obtained if the three selected stress components are chosen realistically. In the calculations, contributions to the results come only from so-called dependence cones.

   This calculation procedure is not based on realistic mechanisms in that there are no mechanisms with which one could actually adjust the three selected, independent stress components in the glacier at will. The dependence cones also have only formal mathematical meaning. No corresponding cone-shaped propagation or influence processes take place in the glacier. Nevertheless, this calculation method yields the correct general solution of the balance and boundary conditions.[17]

9.2 Model Selection

9.2.1 Floating Glaciers

Non-normalized models with stress functions are suitable for floating glaciers. For a floating glacier, the boundary surface Σ of given boundary stresses usually does not allow normalization.[18] Therefore, no alternative models can be considered, neither

[16] The results are given for each of the eight choices "a" to "h" of independent stress components in Sects. 17.1, 17.2, 17.3, 17.4, 17.5, 17.6, 17.7, and 17.8.

[17] "General" solution means ambiguity of the solution. This "generality" or ambiguity reflects the lack of information that prevents the selection of a realistic solution. The basis for this "generality" or ambiguity of the general solution is thus not of a physical nature, but consists in this lack of information, which is why the corresponding calculation procedures also contain formal, non-realistic elements.

[18] See Sect. 7.4.

normalized models with stress functions nor models with three independent stress components.[19]

These non-normalized models with stress functions have a wide range of applications. They can be used to construct the general solution **S** (8.1) of the balance and boundary conditions (2.14)–(2.16) for arbitrarily designed glacier areas Ω[20] and arbitrarily designed boundary surfaces Σ of given boundary stresses. However, this advantage of universal applicability is connected with the disadvantage of redundancies.[21]

9.2.2 Land Glaciers with Multiple Connected Free Surfaces

Normalized models with stress functions are suitable for land glaciers with multiple connected free surfaces Σ,[22] because the normalization requirements[23] can usually be met for land glaciers. Compared to alternative models without normalization, normalized models have the advantage that in the general solution **S** (8.1) the redundancies among the stress functions $\mathbf{A}_0$ are reduced or eliminated. Alternative models with three independent stress components are out of question because of the multiple connection of the free surface Σ.

The normalization requirements can usually be met for land glaciers because the boundary surface Σ of given boundary stresses consists only of the free surface and there is usually a direction that is transverse to this free surface Σ. There may even be two or three mutually perpendicular directions of this kind. To satisfy the normalization requirements, the coordinate system must be rotated so that the corresponding normalization directions are transverse to the boundary surface Σ.[24]

A normalized model with stress functions is applicable for arbitrarily designed glacier domain Ω[25] and the general solution **S** (8.1) of balance and boundary conditions is defined everywhere in Ω if only the free surface Σ is transverse to the normalization directions.[26]

---

[19] In the case of floating glaciers, one could only consider the free surface as the boundary surface Σ of given boundary stresses. In this case, alternative models would be considered, but these would not take into account a reliable piece of information, namely the boundary stresses given by the hydrostatic pressure under water.

[20] Subject to the qualification presupposed in Chap. 2, footnote 1.

[21] See no. 3, Sect. 9.1.1.

[22] See Fig. 7.1c. For example, the free surface is doubly connected if there is a heavy rock on the glacier which breaks the otherwise simple connection.

[23] See point 4, Sect. 9.1.1.

[24] The directions of normalization are tabulated in (7.26) for the different normalizations. The term "transverse" is explained in footnote 25 of Sect. 7.4.

[25] Subject to the qualification presupposed in Chap. 2, footnote 1.

[26] Under this condition, the normalized models with stress functions are suitable for all four cases (a)–(d) shown in Fig. 7.1.

Among the normalized models, the models with xx-yy-xy normalization have[27] the widest range of application, since in this normalization only the z-direction has to be transverse to the free surface Σ.

9.2.3 Land Glaciers with a Simply Connected Free Surface

Models with three independent stress components are suitable for land glaciers with simply connected free surface Σ,[28] as these models have two advantages: No redundancies occur and the variable component $\mathbf{T}_0$ of the general solution $\mathbf{S}$ (8.5) is expressed by the independent stress components, which are less abstract than the components of the $\mathbf{A}_0$-matrix fields in the alternative models with stress functions. The model prerequisites[29] can usually be satisfied for at least one model of type "b", since there is usually a direction that is transverse to the free surface Σ. To satisfy the model prerequisites, the coordinate system must be rotated so that the corresponding model cone is transverse and synchronous with the oriented free surface Σ.[30]

Relatively simple are the models of the types "b", "d" and "e" since in these cases in the calculation of the stress components in each case only integrations occur in one direction, which is expressed by the one-dimensional dependence cones.[31] The models of type "b" have the largest range of application, since in these models[32] the model cone consists of the beam in the positive z-direction and therefore only the positive z-direction must be transverse and synchronous to the oriented free surface Σ.[33] These type "b" models are the simplest in that only integrations in the z-direction occur.

In a model with three independent stress components, the glacier region Ω under consideration is subject to the constraint that it must be compatible[34] with the free surface Σ and the model cone. For example, in a type "b" model, the constrained glacier region Ω is obtained by projecting the free surface Σ in the negative z-direction, since the model cone consists of the ray in the positive z-direction. In the restricted glacier domain Ω, the three independent stress components define both the general solution and the $\mathbf{A}_0$-matrix field of

[27] See Sect. 6.2.2.

[28] See Fig. 7.1a.

[29] See point 1, Sect. 9.1.2.

[30] The model cone and the components $\mathbf{S}_*$ and $\mathbf{T}_0$ of the general solution $\mathbf{S}$ (8.5) are given for each of the eight choices "a" to "h" of independent stress components in Chap. 17. For the respective choices "a" etc. of independent stress components, the special solution $\mathbf{S}_*$ is denoted by $\mathbf{S}_a$ etc..

[31] See Table 8.1, Sect. 8.2.2 and Sects. 17.2, 17.4 and 17.5.

[32] For type "b" and also for the respective other types there are several models, since in each case different orientations of the coordinate system are possible, which lead to different models.

[33] This means that the outward normal of the oriented free surface Σ has a positive z-component.

[34] See Sect. 3.4.1.

the normalized mother model with stress functions, which gives rise to the model with three independent stress components. Therefore, no redundancies occur in this mother model.

Instead of a model with independent stress components and a correspondingly restricted glacier range, one can also use a model with stress functions that is defined in a larger glacier range and can therefore be more advantageous. For example, if instead of a model with independent stress components of type "b" and restricted glacier range Ω,[35] one uses the normalized mother model[36] with normalized stress functions $\mathbf{A}_0$, this mother model provides the general solution $\mathbf{S}$ in any extension[37] of the restricted glacier range Ω. In this mother model, the variable component $\mathbf{T}_0$ of the general solution $\mathbf{S}$ (9.1) is expressed by stress functions $\mathbf{A}_0$ with xx-yy-xy normalization. (9.2) Redundancies may occur outside the restricted glacier domain Ω in that different matrix fields $\mathbf{A}_0$ yield the same component $\mathbf{T}_0$ of the general solution $\mathbf{S}$ despite their normalization.[38]

$$\mathbf{S} = \mathbf{S}_b + \mathbf{T}_0 \tag{9.1}$$

$$\mathbf{T}_0 = \text{rot rot } \mathbf{A}_0; \mathbf{A}_0 \text{ xx} - \text{yy} - \text{xy} - \text{normalized}; \mathbf{A}_{0\Sigma} = \partial_n \mathbf{A}_0 = \mathbf{0} \tag{9.2}$$

---

[35] The restriction of the glacier area arises from the conditions under para. 1c, Sect. 9.1.2. For example, in a model of type "b" the restricted glacier area Ω results from the projection of the free surface Σ in the negative z-direction.

[36] See point 3, Sect. 9.1.2.

[37] Such an extension is subject to the restrictions in Chap. 2, footnote 1.

[38] The solution (9.1) and (9.2) is formally defined in the whole space, since one can also define $\mathbf{A}_0$ in the whole space, and so is $\mathbf{S}_b$ (5.4), since the ice density ρ is zero by definition outside the extended glacier region. However, the solution is only relevant in the extended glacier domain.

Part III

Applications and Examples

10 Land Glaciers

Abstract

For land glaciers with simply connected free surface, the general solution of the balance conditions and the boundary conditions of vanishing boundary stresses on the free surface is expressed by three components of the stress tensor or the deviatoric stress tensor, where these three components can be taken as arbitrary functions within the framework of the general solution. For a glacier with a heavy rock on its surface, such that the free surface surrounding the rock is not simply connected, the general solution of the balance boundary and load conditions is given, where the load conditions are defined by the weight and torque exerted by the rock. For land glaciers with small strain rates, so-called "quasi-stagnant models" are introduced. These quasi-stagnant models are candidates for realistic models and can be represented with reasonable effort without aiming for a precision by too much effort, which cannot be achieved anyway due to information deficits.

In this chapter, some models[1] for the general solution of the balance and boundary conditions (2.14)–(2.16) in land glaciers are discussed. Especially for land glaciers with small strain rates, so-called "quasi-stagnant models" are introduced as candidates for realistic models.

[1] The models are selected according to the selection criteria in Sects. 9.2.2 and 9.2.3.

P. Halfar, *Stresses in glaciers*, https://doi.org/10.1007/978-3-662-66024-9_10

10.1 Glaciers with Simply Connected Free Surface: Models with Three Independent Stress Components

10.1.1 Independent Components S_{xx}, S_{yy}, S_{xy}

In this section, the model type "b" of the general solution[2] with independent stress components S_{xx}, S_{yy} and S_{xy} is discussed. The scope of model type "b" is defined by the following conditions[3]:

- Shape of the free surface Σ and orientation of the coordinate system
  The oriented normal **n** of the free surface Σ must have positive z-components.
  Therefore, the oriented free surface Σ must be transverse and synchronous with the one-dimensional model cone K_z[4] generated by the model cone vector $\mathbf{e}_z$. Σ can therefore be described by a function $z_0(x, y)$ (10.1).
- Glacier area under consideration Ω
  The glacier area Ω considered must be compatible with the model cone K_z and the free surface Σ.[5]
  The glacier area Ω under consideration (10.3) thus lies in the z-direction between the free surface and an area given by a function $z_1(x, y)$.
- Definition domain Ω_{def} of all functions and distributions
  The spatial domain of definition Ω_{def} (10.4) of all used functions and distributions is larger than the glacier area Ω and additionally contains the external area Ω_{ext}, which is generated by all model cones K_z with tip on the free surface Σ. Thus, the external region Ω_{ext} (10.5) lies beyond the free surface Σ in the positive z-direction. In this external region Ω_{ext} all used functions and distributions vanish, especially the independent stress components S_{xx}, S_{yy}, S_{xy} and the ice density ρ.

The general solution **S** (10.6) of the balance and boundary conditions (2.14)–(2.16) and the components $\mathbf{S}_b$ (10.7) and $\mathbf{T}_0$ (10.8) of this solution have the following properties:

1. The selected stress components S_{bxx}, S_{byy} and S_{bxy} of $\mathbf{S}_b$ disappear.
2. The independent stress components T_{0xx}, T_{0yy} and T_{0xy} of $\mathbf{T}_0$ coincide with the arbitrarily chosen independent stress components S_{xx}, S_{yy} and S_{xy} of the general solution **S**.

---

[2] See Table 8.1, column b in Sects. 8.2.2 and 17.2.

[3] See paragraph 1 and paragraph 6, Sect. 9.1.2.

[4] See No. 1, Sect. 3.4.1.

[5] See point 2, Sect. 3.4.1.

3. The balance conditions are fulfilled.
   (a) All stress tensors are symmetrical.
   (b) The divergence of $\mathbf{S}_b$ and of $\mathbf{S}$ is respectively equal to the negative specific ice weight $-\rho\mathbf{g}$ and the divergence of $\mathbf{T}_0$ is zero.
4. The boundary conditions are fulfilled.

At the free surface Σ the boundary stresses of $\mathbf{S}_b$, $\mathbf{T}_0$ and $\mathbf{S}$ are zero.

Some of these properties (numbers 1, 2 and 3a) are obvious. That the balance conditions (number 3b) are fulfilled can be checked with the help of the calculation rules for the differential and integral operators.[6] That according to the boundary conditions (number 4) the boundary stresses of the special solution $\mathbf{S}_b$ vanish follows from the fact that there the special solution $\mathbf{S}_b$ itself vanishes.[7] To prove that the boundary stresses of the weightless stress tensor field $\mathbf{T}_0$ also vanish, one writes these boundary stresses in a form, (10.9) in which differential operators of the type $n_i\partial_k - n_k\partial_i$ acting parallel to the free surface occur. Consequently, the boundary stresses (10.9) of $\mathbf{T}_0$ consist of derivatives parallel to the free surface of functions which vanish at the free surface.[8] Thus these boundary stresses of $\mathbf{T}_0$ vanish.

There is another reason for the disappearance of the boundary stresses of $\mathbf{T}_0$. To do this, consider a surface in the glacier that almost hugs the free surface and its mirror image beyond the free surface. A closed surface in the form of a flat can is created from these two surfaces by connecting the two boundary curves of these surfaces by a narrow annular surface. The bottom of the can consists of the surface in the glacier immediately below the free surface, the can lid consists of the mirror image of this "can bottom" beyond the free surface, and the can wall consists of the narrow annular surface. The weightless stress tensor field $\mathbf{T}_0$ is defined on both sides of the free surface and it produces no force on the closed can surface. Since this weightless stress tensor field $\mathbf{T}_0$ disappears beyond the free surface and thus also on the "can lid", and since the "can wall" can be made arbitrarily narrow so that the force acting on it also disappears in the end, the force on the "can bottom" must also disappear when it is nestled against the free glacier surface from inside the glacier. Since this is true for any such surfaces ("can bottoms"), the boundary stresses of the weightless stress tensor field on the free surface must also disappear.

Integral representations of the general solution are obtained by applying first the differential operators and then the integral operators in the computation of $\mathbf{S}_b$ (10.7) and $\mathbf{T}_0$ (10.8)·[9] Here, one represents the independent stress components S_{xx} etc. and the ice

---

[6] S. (3.21)–(3.27).

[7] The functions $\partial_z^{-1}\rho$, $\partial_x\partial_z^{-2}\rho$ and $\partial_y\partial_z^{-2}\rho$ vanish at the free surface, since ρ by definition vanishes beyond the free surface.

[8] The functions $\partial_z^{-1}S_{xx}$, $\partial_x\partial_z^{-2}S_{xx}$ etc. vanish at the free surface, since S_{xx} etc. vanish by definition beyond the free surface.

[9] The transformations in Sect. 18.5 are used.

density ρ in the glacier region Ω by the smooth independent stress components $\overline{S}_{xx}$ etc. and by the smooth ice density $\overline{\rho}$.[10] In the external region Ω_{ext} beyond the free surface the independent stress components and the ice density vanish by definition. They can be represented with the help of the step function θ, which also vanishes there (10.10)–(10.12).

The three independent stress components S_{xx}, S_{yy} and S_{xy} determine the three complementary stress components S_{xz}, S_{yz} and S_{zz} of the general solution **S** (10.6) by determining the complementary stress components of the weightless stress tensor field $\mathbf{T}_0$ (10.8). The latter (10.15)–(10.17) vanish in the external region Ω_{ext} beyond the free glacier surface and are characterized in the glacier region Ω by the following expressions:

- By functions independent of the ice depth[11] in the z-direction and affected by the independent stress components only by their boundary values at the free surface.
- By functions that depend linearly on the ice depth in the z-direction, vanish at the free surface and are affected by the independent stress components only by their boundary values and by the boundary values of their first derivatives at the free surface.
- By integrals of first derivatives or double integrals of second derivatives of the independent stress components over the ice depth in z-direction.[12]

Thus, the complementary stress components of the general solution **S** depend on the independent stress components and their first and second derivatives on the ice depth in the z-direction. The ray in the positive z-direction thus represents the one-dimensional dependence cone of the general solution with respect to each of the three independent stress components.

The ice density ρ affects the general solution **S** (10.6) by defining the complementary stress components of the stress tensor field $\mathbf{S}_b$ (10.7). These complementary stress components (10.18)–(10.20) of the stress tensor field $\mathbf{S}_b$ vanish in the external region Ω_{ext} beyond the free glacier surface and are characterized in the glacier region by the following expressions:

- By functions that depend linearly on the ice depth in the z-direction, vanish at the free surface, and are affected by the ice density only by its boundary value at the free surface.
- By the integral of the ice density deviation from the surface value over the ice depth in z-direction (If the ice density is spatially constant in the glacier area, this integral disappears).

[10] See No. 5, Sect. 3.1.

[11] In the following, the term "ice depth" stands for both the name and the length of the distance in the z-direction from the free surface to the point considered in the glacier.

[12] The integrals or double integrals are expressions of the type $\partial_z^{-1}(\theta \cdot *)$ or $\partial_z^{-2}(\theta \cdot *)$. See Chap. 3 on integral operators.

- By double integrals of first derivatives of the ice density over the ice depth in z-direction (If the ice density is spatially constant in the glacier area, these double integrals disappear).

Thus, the complementary stress components of the general solution **S** in a point of the considered glacier area depend on the values which the ice density and its first derivatives assume on the ice depth in z-direction. The ray in positive z-direction thus represents the one-dimensional dependence cone of the general solution also with respect to the ice density.

Especially the boundary values (10.21) of the general solution **S** (10.6) at the free surface depend only on the boundary values of the independent stress components. With these boundary values it can be checked again that the boundary stresses disappear (10.2), (10.21), and (10.22).

—

$$\Sigma : z = z_0(x, y) \tag{10.1}$$

$$\mathbf{n} = \left[1 + (\partial_x z_0)^2 + (\partial_y z_0)^2\right]^{-1/2} \cdot \begin{bmatrix} -\partial_x z_0 \\ -\partial_y z_0 \\ 1 \end{bmatrix} \tag{10.2}$$

$$\Omega : z_1(x, y) < z \leq z_0(x, y) \tag{10.3}$$

$$\Omega_{\text{def}} : z_1(x, y) < z < \infty \tag{10.4}$$

$$\Omega_{\text{ext}} : z_0(x, y) < z < \infty \tag{10.5}$$

* * *

$$\mathbf{S} = \mathbf{S}_b + \mathbf{T}_0 \tag{10.6}$$

$$\mathbf{S}_b = \mathbf{S}_b^{\mathrm{T}} = \begin{bmatrix} 0 & 0 & -g_x \partial_z^{-1} \\ 0 & 0 & -g_y \partial_z^{-1} \\ * & * & (g_x \partial_x + g_y \partial_y - g_z \partial_z) \cdot \partial_z^{-2} \end{bmatrix} \rho \tag{10.7}$$

$$\mathbf{T}_0 = \mathbf{T}_0^{\mathrm{T}} = \begin{bmatrix} 1 & 0 & 0 \\ 0 & 0 & 0 \\ * & 0 & \partial_x^2 \partial_z^{-2} \end{bmatrix} S_{xx} + \begin{bmatrix} 0 & 0 & 0 \\ 0 & 1 & -\partial_y \partial_z^{-1} \\ 0 & * & \partial_y^2 \partial_z^{-2} \end{bmatrix} S_{yy} + \begin{bmatrix} 0 & 1 & -\partial_y \partial_z^{-1} \\ 1 & 0 & -\partial_x \partial_z^{-1} \\ * & * & 2\partial_x \partial_y \partial_z^{-2} \end{bmatrix} S_{xy} \tag{10.8}$$

$$
\mathbf{T}_0\mathbf{n} = \begin{bmatrix} (n_x\partial_z - n_z\partial_x)\partial_z^{-1} \\ 0 \\ -(n_x\partial_z - n_z\partial_x)\partial_x\partial_z^{-2} \end{bmatrix} S_{xx}
$$

$$
+ \begin{bmatrix} 0 \\ (n_y\partial_z - n_z\partial_y)\partial_z^{-1} \\ -(n_y\partial_z - n_z\partial_y)\partial_y\partial_z^{-2} \end{bmatrix} S_{yy} + \begin{bmatrix} (n_y\partial_z - n_z\partial_y)\partial_z^{-1} \\ (n_x\partial_z - n_z\partial_x)\partial_z^{-1} \\ -(n_x\partial_z - n_z\partial_x)\partial_y\partial_z^{-2} - (n_y\partial_z - n_z\partial_y)\partial_x\partial_z^{-2} \end{bmatrix} S_{xy} \tag{10.9}
$$

* * *

$$
\theta \overset{\text{def.}}{=} \theta(z_0 - z) \tag{10.10}
$$

$$
S_{ik} \overset{\text{def.}}{=} \theta \cdot \overline{S}_{ik}; \quad (i, k) = (x, x), (y, y), (x, y) \tag{10.11}
$$

$$
\rho \overset{\text{def.}}{=} \theta \cdot \overline{\rho} \tag{10.12}
$$

$$
[\cdot]_0 \overset{\text{def.}}{=} [\cdot]_{z=z_0(x,\, y)} \tag{10.13}
$$

$$
\overline{\rho}_0 \overset{\text{def.}}{=} [\overline{\rho}]_{z=z_0(x,\, y)} \tag{10.14}
$$

$$
\begin{aligned}
T_{0xz} &= -\partial_z^{-1}\partial_x(\theta \cdot \overline{S}_{xx}) - \partial_z^{-1}\partial_y(\theta \cdot \overline{S}_{xy}) \\
&= \theta \cdot \left[\overline{S}_{xx}\right]_0 \cdot \partial_x z_0 - \partial_z^{-1}\left[\theta \cdot \partial_x \overline{S}_{xx}\right] \\
&\quad + \theta \cdot \left[\overline{S}_{xy}\right]_0 \cdot \partial_y z_0 - \partial_z^{-1}\left[\theta \cdot \partial_y \overline{S}_{xy}\right]
\end{aligned} \tag{10.15}
$$

$$
\begin{aligned}
T_{0yz} &= -\partial_z^{-1}\partial_y(\theta \cdot \overline{S}_{yy}) - \partial_z^{-1}\partial_x(\theta \cdot \overline{S}_{xy}) \\
&= \theta \cdot \left[\overline{S}_{yy}\right]_0 \cdot \partial_y z_0 - \partial_z^{-1}\left[\theta \cdot \partial_y \overline{S}_{yy}\right] \\
&\quad + \theta \cdot \left[\overline{S}_{xy}\right]_0 \cdot \partial_x z_0 - \partial_z^{-1}\left[\theta \cdot \partial_x \overline{S}_{xy}\right]
\end{aligned} \tag{10.16}
$$

$$
\begin{aligned}
T_{0zz} &= \partial_z^{-2}\partial_x^2(\theta \cdot \overline{S}_{xx}) + \partial_z^{-2}\partial_y^2(\theta \cdot \overline{S}_{yy}) + 2\partial_z^{-2}\partial_x\partial_y(\theta \cdot \overline{S}_{xy}) \\
&= \theta \cdot \left(^{\partial_x} \cdot \left[_{\overline{S}_{xx}} + (z_0 - z) \cdot \theta \cdot \left\{2 \cdot \partial_x z_0 \cdot \left[\partial_x \overline{S}_{xx}\right]_0 + (\partial_x z_0)^2 \cdot \left[\partial_z \overline{S}_{xx}\right]_0 \right.\right.\right. \\
&+ \partial_x^2 z_0 \cdot \left[\overline{S}_{xx}\right]_0 \Big\} + \partial_z^{-2}\left[\theta \cdot \partial_x^2 \overline{S}_{xx}\right] + \theta \cdot \left(^{\partial_y} \cdot \left[_{\overline{S}_{yy}} + (z_0 - z) \cdot \theta \cdot \left\{2 \cdot \partial_y z_0 \cdot \left[\partial_y \overline{S}_{yy}\right]_0 \right.\right.\right. \\
&+ (\partial_y z_0)^2 \cdot \left[\partial_z \overline{S}_{yy}\right]_0 + \partial_y^2 z_0 \cdot \left[\overline{S}_{yy}\right]_0 \Big\} + \partial_z^{-2}\left[\theta \cdot \partial_y^2 \overline{S}_{yy}\right]
\end{aligned}
$$

$$+2\theta\cdot\partial_x z_0\cdot\partial_y z_0\cdot\Big[\overline{S}_{xy}+2(z_0-z)\cdot\theta\cdot\Big\{\partial_x z_0\cdot\left[\partial_y\overline{S}_{xy}\right]_0+\partial_y z_0\cdot\left[\partial_x\overline{S}_{xy}\right]_0$$
$$+\partial_x z_0\cdot\partial_y z_0\cdot\left[\partial_z\overline{S}_{xy}\right]_0+\partial_x\partial_y z_0\cdot\left[\overline{S}_{xy}\right]_0\Big\}+2\partial_z^{-2}\left[\theta\cdot\partial_x\partial_y\overline{S}_{xy}\right] \tag{10.17}$$

$$\begin{aligned} S_{bxz} &= -g_x\cdot\partial_z^{-1}(\theta\cdot\overline{\rho}) \\ &= g_x\cdot(z_0-z)\cdot\theta\cdot\overline{\rho}_0-g_x\cdot\partial_z^{-1}\{\theta\cdot(\overline{\rho}-\overline{\rho}_0)\} \end{aligned} \tag{10.18}$$

$$\begin{aligned} S_{byz} &= -g_y\cdot\partial_z^{-1}(\theta\cdot\overline{\rho}) \\ &= g_y\cdot(z_0-z)\cdot\theta\cdot\overline{\rho}_0-g_y\cdot\partial_z^{-1}\{\theta\cdot(\overline{\rho}-\overline{\rho}_0)\} \end{aligned} \tag{10.19}$$

$$\begin{aligned} S_{bzz} &= \partial_z^{-2}\cdot\left(g_x\cdot\partial_x+g_y\cdot\partial_y-g_z\cdot\partial_z\right)(\theta\cdot\overline{\rho}) \\ &= \left(g_x\cdot\partial_x z_0+g_y\cdot\partial_y z_0+g_z\right)\cdot(z_0-z)\cdot\theta\cdot\overline{\rho}_0 \\ &\quad -g_z\cdot\partial_z^{-1}\{\theta\cdot(\overline{\rho}-\overline{\rho}_0)\} \\ &\quad +g_x\cdot\partial_z^{-2}(\theta\cdot\partial_x\overline{\rho})+g_y\cdot\partial_z^{-2}(\theta\cdot\partial_y\overline{\rho}) \end{aligned} \tag{10.20}$$

* * *

$$\begin{aligned} [\mathbf{S}]_\Sigma = [\mathbf{T}_0]_\Sigma &= \begin{bmatrix} 1 & 0 & \partial_x z_0 \\ 0 & 0 & 0 \\ \partial_x z_0 & 0 & (\partial_x z_0)^2 \end{bmatrix}_\Sigma\cdot\left[\overline{S}_{xx}\right]_\Sigma \\ &+\begin{bmatrix} 0 & 0 & 0 \\ 0 & 1 & \partial_y z_0 \\ 0 & \partial_y z_0 & (\partial_y z_0)^2 \end{bmatrix}_\Sigma\cdot\left[\overline{S}_{yy}\right]_\Sigma \\ &+\begin{bmatrix} 0 & 1 & \partial_y z_0 \\ 1 & 0 & \partial_x z_0 \\ \partial_y z_0 & \partial_x z_0 & 2\partial_x z_0\cdot\partial_y z_0 \end{bmatrix}_\Sigma\cdot\left[\overline{S}_{xy}\right]_\Sigma \end{aligned} \tag{10.21}$$

$$[\mathbf{S}]_\Sigma\cdot\mathbf{n}=\mathbf{0} \tag{10.22}$$

10.1.2 Independent Non-diagonal Components

In this section, the model type "e" of the general solution[13] with the independent deviatoric stress components S_{xy}, S_{yz}, S_{xz} is discussed. This model type "e" of the general solution is of

[13] See Table 8.1 column e, Sects. 8.2.2 and 17.5.

interest because the independent deviatoric stress components can be determined from measured strain rates of the glacier flow using the flow law. Compared with model type "b", this model type "e" has to fulfil the following, more stringent requirements,[14] which restricts its scope more:

- Shape of the free surface Σ and orientation of the coordinate system
  The outward directed normals of the oriented free surface Σ must have positive x, y and z components.[15]
  Thus the free surface Σ can be represented by a function $x_0(y, z)$ as well as by a function $y_0(x, z)$ as well as by a function $z_0(x, y)$ (10.23). Since these functions all define the same surface Σ, there are relations between these functions which hold near the free surface Σ (10.24) and on Σ (10.25) respectively.
- Glacier area Ω under consideration
  The glacier area Ω considered must be compatible with the model cone K_{xyz} and the free surface.[16]
  This glacier region Ω is thus below the free surface Σ (10.26) with respect to each of the three directions of the coordinate axes, and all cone rays of the model cone K_{xyz} emanating from Ω run uninterruptedly in Ω, until they meet the free surface Σ.
- Definition domain Ω_{def} of all functions and distributions
  The spatial domain of definition Ω_{def} of all used functions and distributions is larger than the glacier area Ω and additionally contains the external area Ω_{ext} beyond the free surface Σ, which is generated by all model cones K_{xyz} with peak on the free surface Σ. In this external region Ω_{ext} beyond the free surface the independent stress components S_{xy}, S_{yz}, S_{xz} and the ice density ρ vanish by definition.

The properties of the general solution **S** (10.27)–(10.29) of model type "e" are the same or similar to the properties of the general solution of model type "b" discussed earlier.[17]

Integral representations of the general solution are obtained by differentiating and then integrating after expressing the ice density ρ and the independent stress components S_{xy} etc. by the step function θ and by smooth functions.$\overline{\rho}$ etc. and $\overline{S}_{xy}$ resp. (10.33), (10.34). With the step function θ, the independent stress components and the ice density in the external region Ω_{ext} beyond the free glacier surface also vanish. The integral representation (10.35)–(10.40) of the general solution **S** (10.27)–(10.29) is simpler compared to model "b" in that no double integrals occur. In contrast to model "b", however, integrations occur not only in the z-direction, but also in the x- and y-directions.

---

[14] See paragraph 1 and paragraph 6, Sect. 9.1.2.

[15] The oriented free surface must be transverse and synchronous to the model cone K_{xyz}, which is generated by the model cone vectors $\mathbf{e}_x$, $\mathbf{e}_y$ and $\mathbf{e}_z$.

[16] See point 2, Sect. 3.4.1.

[17] See Sect. 10.1.1 paragraphs 1–4 and the subsequent discussion of these properties.

For example, the complementary stress component S_{xx} of the general solution **S** (10.27)–(10.29) is affected only by the independent stress components S_{xy} and S_{xz} and the ice density ρ, and is characterized in the glacier domain Ω by expressions of the following types (10.35) and (10.38):

- By functions constant on the ice depth in the x-direction and influenced by the boundary values of the independent stress components on the free surface.
- By integrals over the ice depth in the x-direction with first derivatives of the independent stress components in the integrand.
- By a linear function of the ice depth in the x-direction, which vanishes at the free surface and is influenced by the boundary value of the ice density at the free surface.
- By an integral over the ice depth in x-direction with the deviation of the ice density from its surface value in the integrand (If the ice density is spatially constant in the glacier area, this integral disappears).

Thus, the complementary stress component S_{xx} of the general solution **S** in a point of the considered glacier area depends on the values which the independent stress components as well as their first derivatives and the ice density assume on the ice depth in x-direction. The ray in positive x-direction thus represents the one-dimensional dependence cone of the complementary stress component S_{xx} with respect to the independent stress components and with respect to the ice density. The same applies to the complementary stress components S_{yy} and S_{zz}.

As in model "b", the boundary values (10.41) of the general solution **S** (10.27) at the free surface Σ depend only on the boundary values of the independent stress components.

$$\Sigma : z = z_0(x, y); \quad y = y_0(x, z); \quad x = x_0(y, z) \tag{10.23}$$

$$z \overset{\text{id.}}{=} z_0[x, y_0(x, z)] \overset{\text{id.}}{=} z_0[x_0(y, z), y] \tag{10.24}$$

$$\left.\begin{array}{lcc} \partial_z y_0 & = & 1/\partial_y z_0 \\ \partial_x y_0 & = & -\partial_x z_0/\partial_y z_0 \\ \partial_z x_0 & = & 1/\partial_x z_0 \\ \partial_y x_0 & = & -\partial_y z_0/\partial_x z_0 \end{array}\right\}; \quad \mathbf{r} \in \Sigma \tag{10.25}$$

$$x \leq x_0(y, z); \quad y \leq y_0(x, z); \quad z \leq z_0(x, y); \quad \mathbf{r} \in \Omega \tag{10.26}$$

* * *

$$\mathbf{S} = \mathbf{S}_e + \mathbf{T}_0 \tag{10.27}$$

$$\mathbf{S}_e = \mathbf{S}_e^{\mathrm{T}} = \begin{bmatrix} -g_x \partial_x^{-1} & 0 & 0 \\ 0 & -g_y \partial_y^{-1} & 0 \\ 0 & 0 & -g_z \cdot \partial_z^{-1} \end{bmatrix} \rho \tag{10.28}$$

$$\begin{aligned} \mathbf{T}_0 = \mathbf{T}_0^{\mathrm{T}} = & \begin{bmatrix} -\partial_y \partial_x^{-1} & 1 & 0 \\ 1 & -\partial_x \partial_y^{-1} & 0 \\ 0 & 0 & 0 \end{bmatrix} S_{xy} \\ & + \begin{bmatrix} 0 & 0 & 0 \\ 0 & -\partial_z \partial_y^{-1} & 1 \\ 0 & 1 & -\partial_y \partial_z^{-1} \end{bmatrix} S_{yz} \\ & + \begin{bmatrix} -\partial_z \partial_x^{-1} & 0 & 1 \\ 0 & 0 & 0 \\ 1 & 0 & -\partial_x \partial_z^{-1} \end{bmatrix} S_{xz} \end{aligned} \tag{10.29}$$

* * *

$$\theta \stackrel{\text{def}}{\equiv} \begin{cases} 1 & \text{on } \Omega \\ 0 & \text{otherwise} \end{cases} \tag{10.30}$$

$$\theta = \theta(z_0 - z) = \theta(y_0 - y) = \theta(x_0 - x) \tag{10.31}$$

$$\partial_z \theta = -\delta(z_0 - z) = \partial_z y_0 \cdot \delta(y_0 - y) = \partial_z x_0 \cdot \delta(x_0 - x); \quad (\text{cycl. } x, y, z) \tag{10.32}$$

$$S_{ik} \stackrel{\text{def.}}{=} \theta \cdot \overline{S}_{ik}; \quad (i, k) = (x, y), (y, z), (x, z) \tag{10.33}$$

$$\rho \stackrel{\text{def.}}{=} \theta \cdot \overline{\rho} \tag{10.34}$$

$$\begin{aligned} T_{0xx} &= -\partial_x^{-1} \partial_y (\theta \cdot \overline{S}_{xy}) - \partial_x^{-1} \partial_z (\theta \cdot \overline{S}_{xz}) \\ &= \theta \cdot \left[\overline{S}_{xy}\right]_{x=x_0} \cdot \partial_y x_0 - \partial_x^{-1} \left[\theta \cdot \partial_y \overline{S}_{xy}\right] \\ &\quad + \theta \cdot \left[\overline{S}_{xz}\right]_{x=x_0} \cdot \partial_z x_0 - \partial_x^{-1} \left[\theta \cdot \partial_z \overline{S}_{xz}\right] \end{aligned} \tag{10.35}$$

$$
\begin{aligned}
T_{0yy} &= -\partial_y^{-1}\partial_x\left(\theta\cdot\overline{S}_{xy}\right) - \partial_y^{-1}\partial_z\left(\theta\cdot\overline{S}_{yz}\right) \\
&= \theta\cdot\left[\overline{S}_{xy}\right]_{y=y_0}\cdot\partial_x y_0 - \partial_y^{-1}\left[\theta\cdot\partial_x\overline{S}_{xy}\right] \\
&\quad +\theta\cdot\left[\overline{S}_{yz}\right]_{y=y_0}\cdot\partial_z y_0 - \partial_y^{-1}\left[\theta\cdot\partial_z\overline{S}_{yz}\right]
\end{aligned}
\tag{10.36}
$$

$$
\begin{aligned}
T_{0zz} &= -\partial_z^{-1}\partial_y\left(\theta\cdot\overline{S}_{yz}\right) - \partial_z^{-1}\partial_x\left(\theta\cdot\overline{S}_{xz}\right) \\
&= \theta\cdot\left[\overline{S}_{yz}\right]_{z=z_0}\cdot\partial_y z_0 - \partial_z^{-1}\left[\theta\cdot\partial_y\overline{S}_{zy}\right] \\
&\quad +\theta\cdot\left[\overline{S}_{xz}\right]_{z=z_0}\cdot\partial_x z_0 - \partial_z^{-1}\left[\theta\cdot\partial_x\overline{S}_{xz}\right]
\end{aligned}
\tag{10.37}
$$

$$
\begin{aligned}
S_{exx} &= -g_x\cdot\partial_x^{-1}(\theta\cdot\overline{\rho}) \\
&= g_x\cdot(x_0-x)\cdot\theta\cdot[\overline{\rho}]_{x=x_0} - g_x\cdot\partial_x^{-1}\left\{\theta\cdot\left(\overline{\rho}-[\overline{\rho}]_{x=x_0}\right)\right\}
\end{aligned}
\tag{10.38}
$$

$$
\begin{aligned}
S_{eyy} &= -g_y\cdot\partial_y^{-1}(\theta\cdot\overline{\rho}) \\
&= g_y\cdot(y_0-y)\cdot\theta\cdot[\overline{\rho}]_{y=y_0} - g_y\cdot\partial_y^{-1}\left\{\theta\cdot\left(\overline{\rho}-[\overline{\rho}]_{y=y_0}\right)\right\}
\end{aligned}
\tag{10.39}
$$

$$
\begin{aligned}
S_{ezz} &= -g_z\cdot\partial_z^{-1}(\theta\cdot\overline{\rho}) \\
&= g_z\cdot(z_0-z)\cdot\theta\cdot[\overline{\rho}]_{z=z_0} - g_z\cdot\partial_z^{-1}\left\{\theta\cdot\left(\overline{\rho}-[\overline{\rho}]_{z=z_0}\right)\right\}
\end{aligned}
\tag{10.40}
$$

* * *

$$
\begin{aligned}
[\mathbf{S}]_\Sigma = [\mathbf{T}_0]_\Sigma &= \begin{bmatrix} -\partial_y z_0/\partial_x z_0 & 1 & 0 \\ 0 & -\partial_x z_0/\partial_y z_0 & 0 \\ 0 & 0 & 0 \end{bmatrix}_\Sigma \cdot \left[S_{xy}\right]_\Sigma \\
&+ \begin{bmatrix} 0 & 0 & 0 \\ 0 & 1/\partial_y z_0 & 1 \\ 0 & 1 & \partial_y z_0 \end{bmatrix}_\Sigma \cdot \left[S_{yz}\right]_\Sigma \\
&+ \begin{bmatrix} 1/\partial_x z_0 & 0 & 1 \\ 0 & 0 & 0 \\ 1 & 0 & \partial_x z_0 \end{bmatrix}_\Sigma \cdot [S_{xz}]_\Sigma
\end{aligned}
\tag{10.41}
$$

10.1.3 Independent Deviatoric Components, S'_{xx}, S'_{yy}, S_{xy}

In this section, the model type "f" of the general solution[18] with the independent deviatoric stress components S'_{xx}, S'_{yy}, S_{xy} is discussed. This model type "f" of the general solution is of interest because the independent deviatoric stress components can be determined from measured strain rates of the glacier flow using the flow law. The scope of this model type "f" is defined by the following model assumptions[19]:

- Shape of the free surface Σ and orientation of the coordinate system
  The oriented normals $\mathbf{n}$ of the free surface Σ have positive components with respect to all cone vectors of the rotationally symmetric model cone $K_z^{\odot}$.[20]
  Thus the free surface Σ can be represented by a function $z_0(x, y)$.
- Glacier area Ω under consideration
  The glacier area Ω under consideration must be compatible with the model cone $K_z^{\odot}$ and the free surface.[21] Thus, all cone rays of the model cone originating from Ω run uninterruptedly in Ω, until they meet the free surface Σ.
- Definition domain Ω_{def} of all functions and distributions
  The spatial domain of definition Ω_{def} of all used functions and distributions is larger than the glacier area Ω and additionally contains the external area Ω_{ext} beyond the free surface Σ, which is generated by all model cones $K_z^{\odot}$ with tip on the free surface Σ. In this external region Ω_{ext} beyond the free surface all used functions and distributions vanish, especially the independent deviatoric stress components S'_{xx}, S'_{yy}, S_{xy} and the ice density ρ.[22]

The properties of the general solution $\mathbf{S}$ (10.42)–(10.44) of model type "f" are the same or similar to the properties of the general solution of model type "b" discussed earlier.[23]

Each matrix element of $\mathbf{S}_f$ (10.43) and $\mathbf{T}_0$ (10.44), which constitute the general solution, can be written in distributional form by applying the inverse hyperbolic operator $\Box_z^{-1}$ (3.42) and (3.45) to a distribution (10.50) in which the step function θ (10.46), the delta function δ (10.47) and smooth functions q_1, q_2 and q_3 occur. These functions q_1, q_2 and q_3 are defined by the independent stress components and the ice density, respectively. Since

---

[18] See Table 8.1 column f, Sects. 8.2.2 and 17.6.

[19] See paragraph 1, and paragraph 6, Sect. 9.1.2.

[20] The oriented free surface must be transverse and synchronous to the model cone $K_z^{\odot}$. See para. 1, Sect. 9.1.2.

[21] See point 2, Sect. 3.4.1.

[22] See point 6, Sect. 9.1.2.

[23] See paragraphs 1–4, Sect. 10.1.1 and the subsequent discussion of these properties.

functions q_1, q_2, and q_3 each consist of extensive expressions, they are not listed; only the procedure for their computation is given.[24]

This distributive form (10.50) of a matrix element defines this matrix element as an ordinary function,[25] namely as a sum of three ordinary functions. The value of the first function defined by q_1 at a point $\mathbf{r}$ (10.51) is obtained by integration over the dependency cone $K_z^{\odot}$ starting from $\mathbf{r}$[26], where no contributions come from the cone region beyond the free surface Σ, since the step function θ and hence the integrand vanish there. The value of the second function defined by q_2 at the point $\mathbf{r}$ (10.52) is obtained by integration over the part of the free surface Σ, which lies in this dependence cone $K_z^{\odot}$. The value of the third function defined by q_3 at the point $\mathbf{r}$ (10.53) cannot be written as an integral with an ordinary function in the integrand, but as the z-derivative of a function which is of the same type as the second function. The matrix elements of $\mathbf{S}_f$ each contain no term with q_3 and therefore vanish on the free surface Σ.

As in model "b", the boundary values (10.54) of the general solution $\mathbf{S}$ at the free surface Σ depend only on the boundary values of the independent stress components.[27]

$$\mathbf{S} = \mathbf{S}_f + \mathbf{T}_0 \tag{10.42}$$

$$\mathbf{S}_\mathrm{f} = \mathbf{S}_\mathrm{f}^\mathrm{T} = \begin{bmatrix} \begin{matrix} g_x\partial_x + g_y\partial_y \\ -g_z\partial_z \end{matrix} & 0 & \begin{matrix} \left(g_x\partial_y - g_y\partial_x\right)\partial_y\partial_z^{-1} \\ -g_x\partial_z + g_z\partial_x \end{matrix} \\ 0 & \begin{matrix} g_x\partial_x + g_y\partial_y \\ -g_z\partial_z \end{matrix} & \begin{matrix} \left(g_y\partial_x - g_x\partial_y\right)\partial_x\partial_z^{-1} \\ -g_y\partial_z + g_z\partial_y \end{matrix} \\ * & * & \begin{matrix} g_x\partial_x + g_y\partial_y \\ -g_z\partial_z \end{matrix} \end{bmatrix} \square_z^{-1}\rho \tag{10.43}$$

---

[24]The functions q_1, q_2 and q_3 are calculated for the matrix elements of $\mathbf{T}_0$ from the independent stress components S'_{xx}, S'_{yy}, S_{xy} and for the matrix elements of $\mathbf{S}_f$ from the ice density ρ. The calculation procedure is given in Sect. 18.6. The independent stress components and the ice density are written as the product of the step function θ (10.46) and a smooth function.

[25]The function defined by formula (10.50) is the solution of a hyperbolic differential equation with boundary conditions. S. Chap. 19.

[26]The function $G(\mathbf{r}' - \mathbf{r})$ (10.48) is non-zero only in the points $\mathbf{r}'$ that lie in the dependence cone starting from the point $\mathbf{r}$.

[27]These boundary values (10.54) result from the calculations in Sect. 18.6, where according to formula (19.14) in Chap. 19 only the terms with q_3 play a role.

$$
\mathbf{T}_0 = \mathbf{T}_0^{\mathrm{T}} = \begin{bmatrix} -\partial_y^2 + 2\partial_z^2 & 0 & \partial_x \partial_z^{-1}\left(\partial_y^2 - 2\partial_z^2\right) \\ 0 & \partial_x^2 + \partial_z^2 & -\partial_y \partial_z^{-1}\left(\partial_x^2 + \partial_z^2\right) \\ * & * & 2\partial_x^2 + \partial_y^2 \end{bmatrix} \Box_z^{-1} S'_{xx}
$$

$$
+ \begin{bmatrix} \partial_y^2 + \partial_z^2 & 0 & -\partial_x \partial_z^{-1}\left(\partial_y^2 + \partial_z^2\right) \\ 0 & -\partial_x^2 + 2\partial_z^2 & \partial_y \partial_z^{-1}\left(\partial_x^2 - 2\partial_z^2\right) \\ * & * & \partial_x^2 + 2\partial_y^2 \end{bmatrix} \Box_z^{-1} S'_{yy} \tag{10.44}
$$

$$
+ \begin{bmatrix} 2\partial_x \partial_y & \Box_z & \partial_y \partial_z^{-1} \Box_y \\ * & 2\partial_x \partial_y & \partial_x \partial_z^{-1} \Box_x \\ * & * & 2\partial_x \partial_y \end{bmatrix} \Box_z^{-1} S'_{xy}
$$

* * *

$$
\Sigma : \quad z = z_0(x, y) \tag{10.45}
$$

$$
\theta \stackrel{\text{def.}}{=} \theta[z_0(x, \; y) - z] \tag{10.46}
$$

$$
\delta \stackrel{\text{def.}}{=} \delta[z_0(x, \; y) - z] \tag{10.47}
$$

$$
G(\mathbf{r}' - \mathbf{r}) \stackrel{(3.20)}{=} \frac{1}{2\pi} \cdot \frac{\theta\left[(z' - z) - \sqrt{(x' - x)^2 + (y' - y)^2}\right]}{\sqrt{(z' - z)^2 - (x' - x)^2 - (y' - y)^2}} \tag{10.48}
$$

$$
\mathbf{r} = (x, y, z)^{\mathrm{T}}; \qquad \mathbf{r}' = (x', y', z')^T \tag{10.49}
$$

* * *

$$
\Box_z^{-1}\{\theta \cdot q_1(x, \; y, \; z) + \delta \cdot q_2(x, \; y) + \partial_z[\delta \cdot q_3(x, \; y)]\} \tag{10.50}
$$

$$
\left[\Box_z^{-1}(\theta \cdot q_1)\right](\mathbf{r}) = \int \mathrm{d}x' \mathrm{d}y' \mathrm{d}z' \cdot G(\mathbf{r}' - \mathbf{r}) \cdot \theta(\mathbf{r}') \cdot q_1(\mathbf{r}') \tag{10.51}
$$

$$
\theta(\mathbf{r}') \stackrel{\text{def.}}{=} \theta[z_0(x', \; y') - z']
$$

$$\left[\square_z^{-1}(\delta \cdot q_2)\right](\mathbf{r}) = \int \mathrm{d}x'\mathrm{d}y' \cdot [G(\mathbf{r}' - \mathbf{r})]_{'z_0} \cdot q_2(x', y') \tag{10.52}$$

$$[\cdot]_{'z_0} \overset{\text{def.}}{=} [\cdot]_{z'=z_0(x',\, y')}$$

$$\left[\square_z^{-1}\partial_z(\delta \cdot q_3)\right](\mathbf{r}) =$$

$$\left[\partial_z\square_z^{-1}(\delta \cdot q_3)\right](\mathbf{r}) = \partial_z \int \mathrm{d}x'\mathrm{d}y' \cdot [G(\mathbf{r}' - \mathbf{r})]_{'z_0} \cdot q_3(x', y') \tag{10.53}$$

* * *

$$\begin{aligned}
&[\mathbf{S}]_\Sigma = [\mathbf{T}_0]_\Sigma \\
&= \begin{bmatrix} 2-(\partial_y z_0)^2 & 0 & \partial_x z_0 \cdot \left[2-(\partial_y z_0)^2\right] \\ 0 & 1+(\partial_x z_0)^2 & \partial_y z_0 \cdot \left[1+(\partial_x z_0)^2\right] \\ * & * & 2(\partial_x z_0)^2 + (\partial_y z_0)^2 \end{bmatrix}_\Sigma \cdot \frac{[S'_{xx}]_\Sigma}{N} \\
&+ \begin{bmatrix} 1+(\partial_y z_0)^2 & 0 & \partial_x z_0 \cdot \left[1+(\partial_y z_0)^2\right] \\ 0 & 2-(\partial_x z_0)^2 & \partial_y z_0 \cdot \left[2-(\partial_x z_0)^2\right] \\ * & * & (\partial_x z_0)^2 + 2(\partial_y z_0)^2 \end{bmatrix}_\Sigma \cdot \frac{\left[S'_{yy}\right]_\Sigma}{N} \\
&+ \begin{bmatrix} 2\partial_x z_0 \cdot \partial_y z_0 & N & \partial_y z_0\left[1+(\partial_x z_0)^2-(\partial_y z_0)^2\right] \\ * & 2\partial_x z_0 \cdot \partial_y z_0 & \partial_x z_0\left[1-(\partial_x z_0)^2+(\partial_y z_0)^2\right] \\ * & * & 2 \cdot \partial_x z_0 \cdot \partial_y z_0 \end{bmatrix}_\Sigma \cdot \frac{\left[S'_{xy}\right]_\Sigma}{N}
\end{aligned} \tag{10.54}$$

$$N \overset{\text{def.}}{=} \left[1-(\partial_x z_0)^2-(\partial_y z_0)^2\right]_\Sigma$$

10.2 Glaciers with Surface Load and with Twofold Connected Free Surface: A Model with Normalized Stress Functions

In the following example of a land glacier the simple coherence of its free surface is interrupted by a heavy rock lying on the glacier surface, so that the free glacier surface, which is at the same time the boundary surface Σ of known – namely vanishing – boundary stresses, becomes doubly connected by annularly surrounding the surface Λ_1 loaded by the rock.[28] The glacier surface consisting of the surfaces Σ and Λ_1 is given by a function $z_0(x, y)$ (10.55) and the z-axis of the coordinate system is vertically oriented (10.56) and passes through the loaded surface Λ_1.

This raises the question of the general solution **S** of the balance conditions with vanishing boundary stresses on the free surface Σ, (10.57) which satisfies the load conditions (10.58) and (10.59) on the loaded surface Λ_1. These load conditions mean that the stress tensor field **S** generates on the loaded surface Λ_1 a force $\mathbf{F}_1[\mathbf{S}]$ and a torque $\mathbf{G}_1[\mathbf{S}]$ which are equal to the weight and the torque of the rock. Here, the weight of the rock is determined by its mass m_l and the torque by its mass and the position vector of its centre of gravity$\mathbf{c}_l$.

This general solution **S** (10.60) is composed of three stress tensor fields $\widehat{\mathbf{S}}$, $\mathbf{T}_{**}$ and $\mathbf{T}_0$.[29] The stress tensor field $\widehat{\mathbf{S}}$ (10.67) is a special solution of the balance and boundary conditions (10.61) and on the loaded surface Λ_1 its force and torque (10.62) vanish, because its boundary stresses vanish there. The weightless stress tensor field $\mathbf{T}_{**}$ – constructed below – has vanishing boundary stresses on the free surface (10.63) and on the loaded surface Λ_1 it absorbs the weight and torque of the load (10.64). The general solution (10.60) is obtained by adding any weightless stress tensor fields $\mathbf{T}_0$ whose boundary stresses vanish on the free surface Σ (10.65) and whose forces and torques vanish on the loaded surface Λ_1 (10.66). These stress tensor fields $\mathbf{T}_0$ are given by second derivatives (10.68) of normalized stress functions $\mathbf{A}_0$ (10.69) which are symmetric, which have non-vanishing matrix elements only in their first two rows and columns, and which vanish together with their first derivatives at the free surface Σ of the glacier (10.70). Therefore, their **B**- and **C**-fields (6.8) and (6.10) vanish on Σ and thus all forces and torques (6.12) and (6.13) generated by $\mathbf{T}_0$ on areas inside of any closed paths on Σ vanish also. This implies the above mentioned vanishing of boundary stresses on Σ and vanishing of the force and torque on the surface Λ_1.

The $\mathbf{T}_{**}$ stress tensor field is still missing. It is constructed with the help of the gradient field $\nabla\phi$ (10.71) of the angular coordinate ϕ with respect to the *z-axis*.[30] The integral of this

---

[28] The total boundary surface Λ where no boundary conditions are known consists of two separate and connected parts: the contact surface Λ_1 with the rock and the contact area Λ_0 with the bedrock and the glacier area not under consideration.

[29] This general solution is constructed using the procedure given in Sect. 8.1 according to (8.1). The parameters $\mathbf{F}_1[\mathbf{S}]$ and $\mathbf{G}_1[\mathbf{S}]$ have been assigned their realistic values (10.58) and (10.59).

[30] The stress tensor field $\mathbf{T}_{**}$ is constructed according to the procedure described in Sect. 16.2.

gradient field over closed paths is equal to 2π-times the number of revolutions around the z-axis and, in particular, vanishes for all closed paths which do not include the z-axis. Thus, the rotation of this vector field off the z-axis also vanishes. To remove the undesired singularity of this vector field on the z-axis without changing this vector field in a neighborhood of the annular free surface Σ, one multiplies it by a suitable function $\chi(R)$, which depends only on the distance R from the z-axis. This function $\chi(R)$ disappears in a small, cylindrical environment of the z-axis and has the value 1 (10.73) outside a slightly larger cylindrical environment of the z-axis, so that nothing changes in this cylindrical outer region, in which the annular free surface Σ also lies (10.74).

With the help of this regularized vector field $\chi(R)\,\nabla\,\phi$ a stress function $\mathbf{A}_{**}$ (10.75) with its $\mathbf{B}_{**}$-, $\mathbf{C}_{**}$- and $\mathbf{T}_{**}$-fields (10.78)–(10.80) is defined.[31] The weightless stress tensor field $\mathbf{T}_{**}$ vanishes in the cylindrical outer region mentioned above and thus also on the annular free surface Σ (10.81) and it balances on the surface Λ_1 loaded by the rock the force $\mathbf{F}_1[\mathbf{S}]$ and torque $\mathbf{G}_1[\mathbf{S}]$ (10.82) and (10.83), which have been chosen realistically by the values generated by this rock (10.58) and (10.59). The field $\mathbf{T}_{**}$ (10.80) thus satisfies the balance, boundary and load conditions (10.63) and (10.64).

Thus $\mathbf{S}$ (10.60) is a complete representation of all stress tensor fields satisfying the balance, boundary and load conditions (10.57)–(10.59). This representation consists of the three summands $\widehat{\mathbf{S}}$ (10.67), $\mathbf{T}_{**}$ (10.80) and $\mathbf{T}_0$ (10.68). The summand $\mathbf{T}_0$ is the variable part of this general solution $\mathbf{S}$ and is defined by the second derivatives of three scalar functions A_{0xx}, A_{0yy} and A_{0xy} which vanish together with their first derivatives at the annular free glacier surface Σ (10.70), but which are otherwise arbitrary. Each such triplet of functions leads to a solution $\mathbf{S}$ (10.60) of the balance, boundary and load conditions (10.57)–(10.59) and each solution is representable in this way.

—

$$z = z_0(x, y); \quad \mathbf{r} \in \Sigma \cup \Lambda_1 \tag{10.55}$$

$$\mathbf{g} = -g \cdot \mathbf{e}_z \tag{10.56}$$

* * *

$$\operatorname{div}\mathbf{S} + \rho\mathbf{g} = \mathbf{0}; \quad \mathbf{S} = \mathbf{S}^{\mathrm{T}}; \quad \mathbf{S} \cdot \mathbf{n}|_\Sigma = \mathbf{0} \tag{10.57}$$

$$\int_{\Lambda_1} \mathbf{Sn} \cdot dA \overset{(8.2)}{=} \mathbf{F}_1[\mathbf{S}] \overset{\text{def}}{=} -m_l \cdot g \cdot \mathbf{e}_z \tag{10.58}$$

[31] $\mathbf{A}_{**}$ is constructed according to the general method described by (16.17) in Sect. 16.2.

$$\int_{\Lambda_1} \mathbf{r} \times \mathbf{Sn} \cdot dA \overset{(8.2)}{=} \mathbf{G}_1[\mathbf{S}] \overset{\text{def}}{=} -m_l \cdot \mathrm{g} \cdot \mathbf{c}_l \times \mathbf{e}_z \tag{10.59}$$

* * *

$$\mathbf{S} = \widehat{\mathbf{S}} + \mathbf{T}_{**} + \mathbf{T}_0 \tag{10.60}$$

$$\mathrm{div}\widehat{\mathbf{S}} + \rho \mathbf{g} = 0; \quad \widehat{\mathbf{S}} = \mathbf{S}^{\mathrm{T}}; \quad \mathbf{S} \cdot \mathbf{n}|_{\Sigma} = 0 \tag{10.61}$$

$$\int_{\Lambda_1} \widehat{\mathbf{S}}\mathbf{n} \cdot dA = \mathbf{0}; \quad \int_{\Lambda_1} \mathbf{r} \times \widehat{\mathbf{S}}\mathbf{n} \cdot dA = \mathbf{0} \tag{10.62}$$

$$\mathrm{div}\mathbf{T}_{**} = 0; \quad \mathbf{T}_{**} = \mathbf{T}_{**}^{\mathrm{T}}; \quad \mathbf{T}_{**} \cdot \mathbf{n}|_{\Sigma} = 0 \tag{10.63}$$

$$\int_{\Lambda_1} \mathbf{T}_{**}\mathbf{n} \cdot dA = \mathbf{F}_1[\mathbf{S}]; \quad \int_{\Lambda_1} \mathbf{r} \times \mathbf{T}_{**}\mathbf{n} \cdot dA = \mathbf{G}_1[S] \tag{10.64}$$

$$\mathrm{div}\mathbf{T}_0 = 0; \quad \mathbf{T}_0 = \mathbf{T}_0^{\mathrm{T}}; \quad \mathbf{T}_0 \cdot \mathbf{n}|_{\Sigma} = 0 \tag{10.65}$$

$$\int_{\Lambda_1} \mathbf{T}_0\mathbf{n} \cdot dA = \mathbf{0}; \quad \int_{\Lambda_1} \mathbf{r} \times \mathbf{T}_0\mathbf{n} \cdot dA = \mathbf{0} \tag{10.66}$$

* * *

$$\widehat{\mathbf{S}} = g \cdot \begin{bmatrix} \partial_z^{-1} & 0 & -\partial_x\partial_z^{-2} \\ 0 & \partial_z^{-1} & -\partial_y\partial_z^{-2} \\ -\partial_x\partial_z^{-2} & -\partial_y\partial_z^{-2} & (\partial_x^2+\partial_y^2)\partial_z^{-3}+\partial_z^{-1} \end{bmatrix} \cdot \rho \tag{10.67}$$

* * *

$$\mathbf{T}_0 = \mathbf{T}_0^{\mathrm{T}} = \mathrm{rot\,rot}\mathbf{A}_0 = \begin{bmatrix} \partial_z^2 A_{0yy} & -\partial_z^2 A_{0xy} & -\partial_x\partial_z A_{0yy}+\partial_y\partial_z A_{0xy} \\ * & \partial_z^2 A_{0xx} & -\partial_y\partial_z A_{0xx}+\partial_x\partial_z A_{0xy} \\ * & * & \partial_x^2 A_{0yy}+\partial_y^2 A_{0xx}-2\partial_x\partial_y A_{0xy} \end{bmatrix} \tag{10.68}$$

$$\mathbf{A}_0 = \mathbf{A}_0^{\mathrm{T}} = \begin{bmatrix} A_{0xx} & A_{0xy} & 0 \\ * & A_{0yy} & 0 \\ 0 & 0 & 0 \end{bmatrix} \tag{10.69}$$

$$\mathbf{A}_0|_{\Sigma} = \partial_z \mathbf{A}_0|_{\Sigma} = \mathbf{0} \tag{10.70}$$

* * *

$$\nabla\varphi = \frac{1}{R^2}\cdot(\mathbf{e}_z\times\mathbf{r}) = \frac{1}{R^2}\cdot(\mathbf{e}_z\times\mathbf{R}) \tag{10.71}$$

$$\mathbf{r} = x\cdot\mathbf{e}_x + y\cdot\mathbf{e}_y + z\cdot\mathbf{e}_z;\quad \mathbf{R} = x\cdot\mathbf{e}_x + y\cdot\mathbf{e}_y;\quad R = \sqrt{x^2+y^2} \tag{10.72}$$

$$R_0 \overset{\text{prec.}}{<} R_1 : \chi(R) = \begin{cases} 0; & R \overset{\text{prec.}}{<} R_0 \\ 1; & R_1 \overset{\text{prec.}}{<} R \end{cases} \tag{10.73}$$

$$R_1 < R;\quad \boldsymbol{r} \overset{\text{prec.}}{\in} \Sigma \tag{10.74}$$

$$\mathbf{A}_{**} \overset{\text{def}}{=} \underbrace{[-m_l\cdot \mathrm{g}\cdot\mathbf{e}_z\times(\mathbf{r}-\mathbf{c}_1)]}_{\mathbf{G}_1[\mathbf{S}]+\mathbf{F}_1[\mathbf{S}]\times\mathbf{r}}\cdot\underbrace{\frac{\chi(R)}{2\pi R^2}\cdot(\mathbf{e}_z\times\mathbf{r})^{\mathrm{T}}}_{\chi(R)\cdot\nabla^{\mathrm{T}}\varphi/(2\pi)} \tag{10.75}$$

$$\operatorname{rot}\mathbf{A}_{**} = -\frac{m_l\cdot \mathrm{g}\cdot\chi(R)}{2\pi\cdot R^2}\cdot\mathbf{e}_z(\mathbf{e}_z\times\mathbf{r})^{\mathrm{T}} - \frac{\chi'(R)\cdot m_l\cdot \mathrm{g}}{2\pi R}\cdot\mathbf{e}_z\cdot[\mathbf{e}_z\times(\mathbf{r}-\mathbf{c}_l)]^{\mathrm{T}} \tag{10.76}$$

$$\operatorname{rot}\operatorname{rot}\mathbf{A}_{**} = -\frac{m_l\cdot \mathrm{g}}{2\pi R}\{\chi'(R) + [R\cdot\chi'(R)]' + \frac{\mathbf{c}_l\cdot\mathbf{R}}{R^2}\cdot[2\chi'(R) - [R\cdot\chi'(R)]']\}\cdot\mathbf{e}_z\cdot\mathbf{e}_z^{\mathrm{T}} \tag{10.77}$$

$$\mathbf{B}_{**} = \underbrace{-m_l\cdot \mathrm{g}\cdot\mathbf{e}_z}_{\mathbf{F}_1[\mathbf{S}]}\cdot\underbrace{\frac{\chi(R)}{2\pi R^2}\cdot(\mathbf{e}_z\times\mathbf{r})^{\mathrm{T}}}_{\chi(R)\cdot\nabla^{\mathrm{T}}\varphi/(2\pi)} - \frac{m_l\cdot \mathrm{g}\cdot\chi'(R)}{2\pi R}\cdot\mathbf{e}_z\cdot[\mathbf{e}_z\times(\mathbf{r}-\mathbf{c}_1)]^{\mathrm{T}} \tag{10.78}$$

$$\mathbf{C}_{**} = \underbrace{m_l\cdot \mathrm{g}\cdot(\mathbf{e}_z\times\mathbf{c}_1)}_{\mathbf{G}_1[\mathbf{S}]}\cdot\underbrace{\frac{\chi(R)}{2\pi R^2}\cdot(\mathbf{e}_z\times\mathbf{r})^{\mathrm{T}}}_{\chi(R)\cdot\nabla^{\mathrm{T}}\phi/(2\pi)} + \frac{m_l\cdot \mathrm{g}\cdot\chi'(R)}{2\pi R}\cdot(\mathbf{e}_z\times\mathbf{r})$$
$$\cdot[\mathbf{e}_z\times(\mathbf{r}-\mathbf{c}_1)]^{\mathrm{T}} \tag{10.79}$$

$$\mathbf{T}_{**} = -\frac{m_l\cdot \mathrm{g}}{2\pi R}\left\{\chi'(R) + [R\cdot\chi'(R)]' + \frac{\mathbf{c}_l\cdot\mathbf{R}}{R^2}\cdot\left[2\chi'(R) - [R\cdot\chi'(R)]'\right]\right\}\cdot\mathbf{e}_z\mathbf{e}_z^{\mathrm{T}} \tag{10.80}$$

$$\mathbf{T}_{**}|_{\Sigma} \overset{(10.80),(10.73),(10.74)}{=} \mathbf{0} \tag{10.81}$$

$$\int_{\Lambda_1}\mathbf{T}_{**}\mathbf{n}\cdot dA \overset{(6.12)}{=} \oint_{\partial\Lambda_1}\mathbf{B}_{**}\cdot d\mathbf{r} \overset{(10.78),(10.73)}{=} \mathbf{F}_1[\mathbf{S}] \tag{10.82}$$

$$\int_{\Lambda_1}\mathbf{r}\times\mathbf{T}_{**}\mathbf{n}\cdot dA \overset{(6.13)}{=} \oint_{\partial\Lambda_1}\mathbf{C}_{**}\cdot d\mathbf{r} \overset{(10.79),(10.73)}{=} \mathbf{G}_1[\mathbf{S}] \tag{10.83}$$

10.3 Quasi-Stagnant Models

For quasi-stagnant glaciers, i.e. glaciers with very small strain rates, whose stress tensor fields differ only slightly from a stagnant stress tensor fiel,[32] so-called quasi-stagnant stress tensor fields are introduced. These quasi-stagnant stress tensor fields solve the task of finding candidates for realistic solutions with reasonable computational effort, without aiming at a precision by too high computational effort, which can never be achieved due to unavoidable information deficits.

10.3.1 Stagnant Glaciers

In this section the stagnant stress tensor fields $\breve{\mathbf{S}}$[33] are described. They are intended to serve as references for stress tensor fields in quasi-stagnant glaciers.

A stagnant, that is, motionless glacier resembles a liquid at rest in a vessel.[34] Thus, a stagnant glacier lies in a trough, has a horizontal free surface Σ with vertically upward oriented normal $\breve{\mathbf{n}}$ (10.84), and its stagnant stress tensor field $\breve{\mathbf{S}}\breve{\mathbf{S}}$ (10.85) is isotropic and defined by a pressure field $\breve{p}$. From the balance condition (10.86) it follows that both the pressure $\breve{p}$ and the ice density $\breve{\rho}$ are horizontally homogeneous (10.88) and (10.89). At the free surface, the boundary stresses vanish and so does the pressure. (10.87) At a point **r,** this pressure $\breve{p}(\boldsymbol{r})$ is given $\breve{p}(\boldsymbol{r})$ by the path integral (10.90) of the pressure gradient (10.86), where the integration path starts at free surface Σ, ends at the point, **r** and is otherwise arbitrary. One can also calculate the pressure field $\breve{p}$ (10.92) by means of an integral operator $(\mathbf{a}\nabla)^{-1}$.[35]

—

[32] The term "stagnant stress tensor field" is a shorthand term for the stress tensor field of a stagnant glacier. Stagnant stress tensor fields have vanishing deviatoric components and are therefore scalar multiples of the unit tensor.

[33] In the stagnant case, all symbols are marked with the accent "˘".

[34] Despite this comparison with the hydrostatic case, we do not use the seemingly obvious term "glaciostatic" here. In fact, all models in this paper are glaciostatic, since there is complete balance of all forces and all torques. The special case of stagnant glaciers is thus not characterized by glaciostatics, but by the isotropy of the stress tensor fields, which is equivalent to deviatoric stress components disappearing everywhere.

[35] The vector **a** must be transverse to the free surface (see footnote 25, Sect. 7.4) and the integration cone of $(\mathbf{a}\nabla)^{-1}$ must be directed upwards (see footnote 11, Sect. 3.2).

$$\breve{\mathbf{n}} = -\frac{\mathbf{g}}{g} \tag{10.84}$$

$$\breve{\mathbf{S}} = -\breve{p} \cdot \mathbf{1} \tag{10.85}$$

$$\nabla\breve{p} = \breve{\rho} \cdot \mathbf{g} \tag{10.86}$$

$$\breve{p}|_{\breve{\Sigma}} = 0 \tag{10.87}$$

* * *

$$\nabla\breve{p} \times \mathbf{g} = \mathbf{0} \tag{10.88}$$

$$\nabla\breve{p} \times \mathbf{g} = \mathbf{0} \tag{10.89}$$

$$\breve{p}(\boldsymbol{r}) = \int_{\breve{\Sigma}}^{r} \breve{\rho}(\boldsymbol{r}') \cdot \mathbf{g} \cdot \mathrm{d}\mathbf{r}' = -\int_{\mathbf{r}}^{\breve{\Sigma}} \breve{\rho}(\mathbf{r}') \cdot \mathbf{g} \cdot \mathrm{d}\mathbf{r}' \tag{10.90}$$

$$(\mathbf{a}\nabla)\breve{p} = \mathbf{a}\mathbf{g} \cdot \breve{\rho} \tag{10.91}$$

$$\breve{p} = (\mathbf{a}\mathbf{g}) \cdot (\mathbf{a}\nabla)^{-1}\breve{\rho} \tag{10.92}$$

10.3.2 Quasi-Stagnant Models

There are glaciers whose stress distribution differs little from a stagnant stress distribution (10.85). As candidates for the description of these stress distributions the quasi-stagnant stress tensor fields are introduced.

A quasi-stagnant stress tensor field is defined in the context of the general solution of the balance and boundary conditions (2.14)–(2.16) by additional conditions of the following kind:

- The additional conditions are also fulfilled by the stagnant stress tensor fields $\breve{\mathbf{S}}$ (10.85).
- The additional conditions uniquely define the quasi-stagnant stress tensor field.

Consequently, a quasi-stagnant stress tensor field must change into a stagnant stress tensor field $\breve{\mathbf{S}}$ (10.85), if the considered model glacier is changed into a trough glacier with horizontally homogeneous ice density and horizontal free surface, because in such a trough glacier the rigid stress tensor field $\breve{\mathbf{S}}$ (10.85) is the unique solution of the balance and boundary conditions (2.14)–(2.16) as well as the additional conditions. Thus, in the ideal case of a stagnant trough glacier, a quasi-stagnant stress tensor field coincides with the stagnant stress tensor field $\breve{\mathbf{S}}$, which in this case is also the actual stress tensor field. It is therefore obvious to consider a quasi-stagnant stress tensor field also in a quasi stagnant – i.e. almost stagnant – glacier as a way to describe the actual stress tensor field. Whether this quasi-stagnant stress tensor field then corresponds sufficiently accurately to the actual stress tensor field cannot be assessed within the framework of this general paper.[36]

There are infinitely many quasi-stagnant stress tensor fields. Thus, one can start from the unique representation of the general solution $\mathbf{S}$ by three independent stress components[37] and construct quasistagnant stress tensor fields from them by specifying these three independent stress components by conditions which are also valid in the stagnant case. There are an immense number of possibilities for this method. For example, at points $\mathbf{r}$ in the glacier one can express the diagonal ones among the three independent stress components by a path integral (10.93) running over a path from the respective point $\mathbf{r}$ to the free ice surface Σ, and one can express the deviatoric ones among the three independent stress components by the difference of two such path integrals. The stress tensor field thus defined is quasi-stagnant, since in the horizontally homogeneous case it resembles the rigid stress tensor field (10.85) and (10.90). One could make it even more sophisticated by averaging over several such path integrals for the diagonal stress components and over several such differences of path integrals for the deviatoric stress components.

So the quasi-stagnant stress tensor fields are merely candidates for passable approximations to actual stress tensor fields and the construction of quasi-stagnant stress tensor fields is not a well-defined mathematical procedure but a heuristic method to find passable stress tensor fields. In this search for passable quasi-stagnant stress tensor fields, one can first consider the simple cases.[38] For example, $\mathbf{S}_e$, $\mathbf{S}_f$, $\mathbf{S}_g$, and $\mathbf{S}_h$ are relatively simple quasi-stagnant stress tensor fields defined by the fact that their respective three selected deviatoric stress components vanish.[39]

$$\int_{\mathbf{r}}^{\Sigma} \rho(\mathbf{r}') \cdot \mathbf{g} \cdot d\mathbf{r}' \tag{10.93}$$

---

[36] This can only be decided after further investigation in each specific case.

[37] See Sect. 8.2.

[38] "Simple" means that the quasi-stagnant stress tensor field can be represented mathematically in a relatively simple way.

[39] See formulae (17.41), (17.49), (17.59) and (17.71).

10.3.3 The Quasi-Stagnant Model with Horizontally Acting Gravity Pressure

In order to present a particularly simple example of a quasi-stagnant stress tensor field, the quasi-stagnant stress tensor field **S** with horizontally acting gravity pressure is examined in this section. The gravity pressure p (10.94) at a point **r** is by definition equal to the fictitious pressure that a very narrow column of ice of the glacier cut out vertically above this point **r** would exert by its weight on its horizontal base.

In the following, the z-axis *is* oriented vertically upward (10.95), the free glacier surface Σ is defined by a function $z_0(x, y)$ (10.96), and the ice density ρ (10.97) in the glacier is given by a smooth function $\overline{\rho}$ and vanishes above the free surface Σ, which is caused by the step function θ also vanishing there. The quasi-rigid stress tensor field $\widehat{\mathbf{S}}$ (10.99) with horizontally acting gravity pressure is obtained from the general solution **S** (10.6) with independent horizontal stresses S_{xx}, S_{yy} and S_{xy} by replacing the longitudinal stresses S_{xx} and S_{yy} by the negative gravity pressure p (10.98) and letting the shear stress S_{xy} vanish.

If the ice density in the glacier has a spatially constant value ρ_c, everything becomes simpler (10.100)–(10.103). In this case, the quasi-stagnant stress tensor field $\widehat{\mathbf{S}}$ can also be represented using the following geometric quantities:

- Ice depth $z_0 - z$
- Inclinations $\tan\alpha_x$ and $\tan\alpha_y$ of the glacier surface in x- and y- direction, and amount $|\tan \alpha_{\max}|$ of maximum inclination of the glacier surface (10.104)–(10.106)
- Radii of curvature R_x and R_y of the glacier surface in x- and y-direction (10.107) and (10.108).

This quasi-stagnant stress tensor field $\widehat{\mathbf{S}}$ (10.110) is obviously almost stagnant only if the surface slope angles are small and the ice depth is small compared to the radii of curvature (10.111), which is assumed in the following.

The quasi-rigid stress tensor field $\widehat{\mathbf{S}}$ (10.110) is best characterized by the stress vectors acting on fictitious vertical planes with their consequently horizontal normal vectors **h** (10.112). The normal components of these stress vectors $\widehat{\mathbf{S}}\mathbf{h}$ (10.113) are equal to the negative gravity pressure p (10.102). The shear components are vertically directed and are such that these stress vectors $\widehat{\mathbf{S}}\mathbf{h}$ are parallel to the ice surface (Fig. 10.1). Therefore, the direction of the contour line through a point on the glacier surface is a principal stress direction of the quasi-stagnant stress tensor field $\widehat{\mathbf{S}}$ (10.110) in all vertically underlying points, with negative gravity pressure p as the principal stress.[40]

To calculate the stresses at the glacier bed Σ_1, the bed is defined by a function $z_1(x, y)$, and a positively oriented orthonormal basis $\mathbf{n}_1, \mathbf{l}_1, \mathbf{m}_1$ is introduced at each point on the bed. This orthonormal basis consists of the oriented normal vector $\mathbf{n}_1$ (10.124), the tangential

[40] In this case, the surface slope $\tan\alpha$ in formula (10.113) disappears.

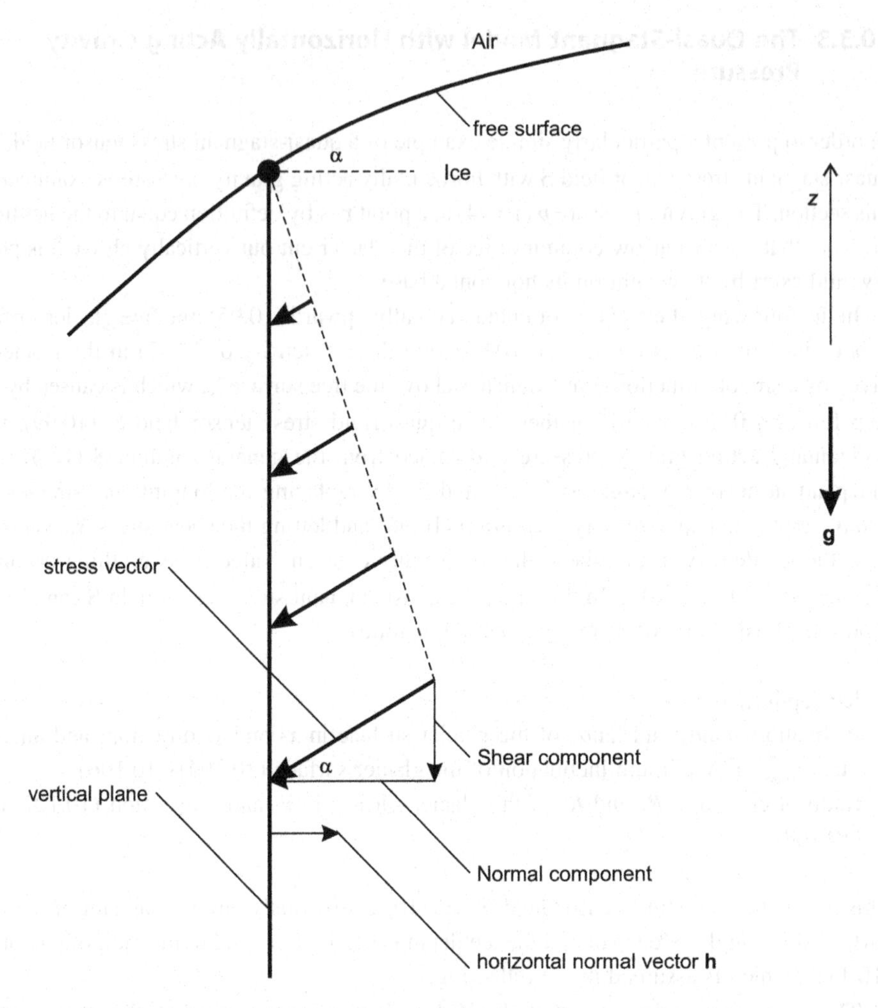

Fig. 10.1 Stress vectors of the quasi-stagnant stress tensor field $\widehat{\mathbf{S}}$ on a fictitious vertical plane in the glacier. The figure shows a vertical section perpendicular to this plane. The stress vectors lie in the section plane, are parallel to the glacier surface and vary linearly with ice depth

unit vector $\mathbf{l}_1$ (10.126) in the direction of the bed slope, and the horizontal tangential unit vector $\mathbf{m}_1$ (10.127) parallel to the contour line of the bed. At the points on the glacier surface, an orthonormal basis $\mathbf{n}_0$, $\mathbf{l}_0$, $\mathbf{m}_0$ is introduced in the same way (10.115)–(10.118). In addition, the tangential unit vector $\mathbf{l}'_0$ (10.129) is defined at the glacier bed, which has the same horizontal orientation as the vector $\mathbf{l}_0$ in the direction of the surface slope.[41]

[41] This tangential vector $\mathbf{l}'_0$ at the bed is obtained by projecting the tangential vector $\mathbf{l}_0$ of the ice surface, pointing in the direction of the surface slope, vertically onto the tangential plane of the bed and then normalizing it to length 1.

The stress $\widehat{\mathbf{S}}\mathbf{n}_1$ acting on the bedrock at the base of the glacier can be expanded in the normal vector $\mathbf{n}_1$ (normal stress) and the oblique-angled tangential vectors $\mathbf{l}_1$ and $\mathbf{l}'_0$ (shear stress). (10.132) The following quantities occur:

- Gravity pressure p_1 (10,130) at the bottom of the glacier.
- Amount $|\tan \alpha_{max}|$ (10.106) of surface slope and amount $|\tan \beta_{max}|$ (10.123) of bottom slope.
- Scalar product of unit vectors $\mathbf{m}_1$ (10.127) and $\mathbf{m}_0$ (10.118), which are parallel to the contour lines at the subsurface and ice surface, respectively.
- The small quantity ϵ (10.109), which is the approximate average of the two quotients between the ice depth and the two radii of curvature of the ice surface, according to Prerequisite (10.111).

If the slope of the glacier bed is not too great and under the conditions mentioned (10.111), the shear component of the subsurface stress $\widehat{\mathbf{S}}\mathbf{n}_1$ (10.132) is dominated by the $\mathbf{l}'_0$-term. Thus, in this case, the result is that, to a good approximation, the shear stress at the glacier bed is determined by the slope of the glacier surface. The shear stress acting on the bedrock points in the same horizontal direction as the surface slope and the magnitude of this shear stress is obtained in good approximation from the gravity pressure p_1 at the bed by multiplying it by the magnitude $|\tan \alpha_{max}|$ of the surface slope [3, p. 104; 4, p. 241].

$$p(\mathbf{r}) = \mathrm{g}^2 \cdot (\mathbf{g}\nabla)^{-1}\rho = \mathrm{g}^2 \cdot \int_0^\infty d\alpha \cdot \rho(\mathbf{r} - \alpha\mathbf{g}) = \mathrm{g} \cdot \int_0^\infty d\alpha \cdot \rho(\mathbf{r} - \frac{\mathbf{g}}{\mathrm{g}} \cdot \alpha) \tag{10.94}$$

$$\mathbf{g} = -g \cdot \mathbf{e}_z \tag{10.95}$$

$$z = z_0(x, y); \quad \mathbf{r} \overset{\text{prec.}}{\in} \Sigma \tag{10.96}$$

$$\rho = \theta(z_0 - z) \cdot \overline{\rho} \tag{10.97}$$

$$p(x, y, z) = -g \cdot \partial_z^{-1}\rho = g \cdot \int_0^\infty d\alpha \cdot \rho(x, y, z + \alpha) \tag{10.98}$$

$$\widehat{\mathbf{S}} = \begin{bmatrix} -1 & 0 & \partial_x\partial_z^{-1} \\ 0 & -1 & \partial_y\partial_z^{-1} \\ * & * & -(\partial_x^2+\partial_y^2)\partial_z^{-2}-1 \end{bmatrix} \cdot p = \begin{bmatrix} \partial_z^{-1} & 0 & -\partial_x\partial_z^{-2} \\ 0 & \partial_z^{-1} & -\partial_y\partial_z^{-2} \\ * & * & (\partial_x^2+\partial_y^2)\partial_z^{-3}+\partial_z^{-1} \end{bmatrix} \cdot \rho g \tag{10.99}$$

* * *

$$\overline{\rho} = \rho_c; \quad \nabla \rho_c = \mathbf{0} \tag{10.100}$$

$$\partial_z^{-n} \rho = \Theta(z_0 - z) \cdot \rho_c \cdot \frac{(-1)^n (z_0 - z)^n}{n!}; \quad n = 0,1,\ldots \tag{10.101}$$

$$p = \Theta(z_0 - z) \cdot g \cdot \rho_c \cdot (z_0 - z) \tag{10.102}$$

$$\widehat{\mathbf{S}} = \widehat{\mathbf{S}}^{\mathrm{T}} = -p \begin{bmatrix} 1 & 0 & \partial_x z_0 \\ 0 & 1 & \partial_y z_0 \\ * & * & \begin{matrix} 1+(\partial_x z_0)^2+(\partial_y z_0)^2 \\ +(z_0-z)\left(\partial_x^2+\partial_y^2\right) z_0/2 \end{matrix} \end{bmatrix} \tag{10.103}$$

* * *

$$\tan \alpha_x = \partial_x z_0 \tag{10.104}$$

$$\tan \alpha_y = \partial_y z_0 \tag{10.105}$$

$$|\tan \alpha_{\max}| = \sqrt{(\partial_x z_0)^2 + (\partial_y z_0)^2} \tag{10.106}$$

$$R_x = \frac{\left[1 + (\partial_x z_0)^2\right]^{3/2}}{\partial_x^2 z_0} = \frac{1}{\partial_x^2 z_0 \cdot \cos^3 \alpha_x} \tag{10.107}$$

$$R_y = \frac{\left[1 + (\partial_y z_0)^2\right]^{3/2}}{\partial_y^2 z_0} = \frac{1}{\partial_y^2 z_0 \cdot \cos^3 \alpha_y} \tag{10.108}$$

$$\begin{aligned} \varepsilon &\overset{\text{def.}}{=} \frac{1}{2} \cdot (z_0 - z)\left(\partial_x^2 + \partial_y^2\right) z_0 \\ &= \frac{1}{2} \cdot \left[\frac{z_0 - z}{R_x \cdot \cos^3 \alpha_x} + \frac{z_0 - z}{R_y \cdot \cos^3 \alpha_y}\right] \end{aligned} \tag{10.109}$$

$$\widehat{\mathbf{S}} = \widehat{\mathbf{S}}^{\mathrm{T}} = -p \cdot \begin{bmatrix} 1 & 0 & \tan \alpha_x \\ 0 & 1 & \tan \alpha_y \\ * & * & 1 + \tan^2 \alpha_{\max} + \varepsilon \end{bmatrix} \tag{10.110}$$

$$|\tan \alpha_{\max}|, \left|\frac{z_0 - z}{R_x}\right|, \left|\frac{z_0 - z}{R_y}\right| \overset{\text{prec.}}{\ll} 1 \tag{10.111}$$

* * *

$$\mathbf{h} \overset{\text{def.}}{=} \begin{bmatrix} h_x \\ h_y \\ 0 \end{bmatrix}; \qquad |\,\mathbf{h}\,| = \sqrt{h_x^2 + h_y^2} \overset{\text{prec.}}{=} 1 \tag{10.112}$$

$$\widehat{\mathbf{S}}\mathbf{h} = -p \cdot (\mathbf{h} + \tan\alpha \cdot \mathbf{e}_z) \tag{10.113}$$

$$\tan\alpha = h_x \cdot \partial_x z_0 + h_y \cdot \partial_y z_0 \tag{10.114}$$

* * *

$$\mathbf{n}_0 = \frac{1}{N_0} \begin{bmatrix} -\partial_x z_0 \\ -\partial_y z_0 \\ 1 \end{bmatrix}, \qquad |\,\mathbf{n}_0\,| = 1 \tag{10.115}$$

$$N_0 = \sqrt{1 + (\partial_x z_0)^2 + \left(\partial_y z_0\right)^2} = \sqrt{1 + \tan^2\alpha_{\max}} = \frac{1}{\cos\alpha_{\max}} \tag{10.116}$$

$$\mathbf{l}_0 = \frac{-1}{M_0 N_0} \cdot \begin{bmatrix} \partial_x z_0 \\ \partial_y z_0 \\ (\partial_x z_0)^2 + \left(\partial_y z_0\right)^2 \end{bmatrix}; \qquad |\,\mathbf{l}_0\,| = 1 \tag{10.117}$$

$$\mathbf{m}_0 = \mathbf{n}_0 \times \mathbf{l}_0 = \frac{1}{M_0} \cdot \begin{bmatrix} \partial_y z_0 \\ -\partial_x z_0 \\ 0 \end{bmatrix}; \qquad |\,\mathbf{m}_0\,| = 1 \tag{10.118}$$

$$M_0 = \sqrt{(\partial_x z_0)^2 + \left(\partial_y z_0\right)^2} = |\,\tan\alpha_{\max}\,| \tag{10.119}$$

* * *

$$z = z_1(x, y); \qquad \mathbf{r} \overset{\text{prec.}}{\in} \Sigma_1 \tag{10.120}$$

$$\tan\beta_x = \partial_x z_1 \tag{10.121}$$

$$\tan\beta_y = \partial_y z_1 \tag{10.122}$$

$$|\,\tan\beta_{\max}\,| = \sqrt{(\partial_x z_1)^2 + \left(\partial_y z_1\right)^2} \tag{10.123}$$

$$\mathbf{n}_1 = \frac{1}{N_1} \cdot \begin{bmatrix} -\partial_x z_1 \\ -\partial_y z_1 \\ 1 \end{bmatrix}; \quad |\, \mathbf{n}_1 \,| = 1 \tag{10.124}$$

$$N_1 = \sqrt{1 + (\partial_x z_1)^2 + (\partial_y z_1)^2} = \sqrt{1 + \tan^2\beta_{\max}} = \frac{1}{\cos\beta_{\max}} \tag{10.125}$$

$$\mathbf{l}_1 = \frac{-1}{M_1 N_1} \cdot \begin{bmatrix} \partial_x z_1 \\ \partial_y z_1 \\ (\partial_x z_1)^2 + (\partial_y z_1)^2 \end{bmatrix}; \quad |\, \mathbf{l}_1 \,| = 1 \tag{10.126}$$

$$\mathbf{m}_1 = \mathbf{n}_1 \times \mathbf{l}_1 = \frac{1}{M_1} \cdot \begin{bmatrix} \partial_y z_1 \\ -\partial_x z_1 \\ 0 \end{bmatrix}; \quad |\, \mathbf{m}_1 \,| = 1 \tag{10.127}$$

$$M_1 = \sqrt{(\partial_x z_1)^2 + (\partial_y z_1)^2} = |\, \tan\beta_{\max} \,| \tag{10.128}$$

* * *

$$\mathbf{l}_0' = \frac{-1}{M_0 \cdot \sqrt{1 + M_1^2 \cdot (\mathbf{m}_0 \mathbf{m}_1)^2}} \cdot \begin{bmatrix} \partial_x z_0 \\ \partial_y z_0 \\ \partial_x z_0 \cdot \partial_x z_1 + \partial_y z_0 \cdot \partial_y z_1 \end{bmatrix}; \quad |\, \mathbf{l}_0' \,| = 1 \tag{10.129}$$

$$p_1 = g \cdot \rho_c \cdot (z_0 - z_1) \tag{10.130}$$

$$\partial_x z_0 \cdot \partial_x z_1 + \partial_y z_0 \cdot \partial_y z_1 = |\, \tan\alpha_{\max} \,| \cdot |\, \tan\beta_{\max} \,| \cdot \mathbf{m}_0 \mathbf{m}_1 \tag{10.131}$$

* * *

$$\begin{aligned} &\widehat{\mathbf{S}} \cdot \mathbf{n}_1 \\ &= -\mathbf{n}_1 \cdot p_1 \left\{ 1 + \cos^2\beta_{\max} \cdot \left[\tan^2\alpha_{\max} - 2 |\tan\alpha_{\max} \cdot \tan\beta_{\max}| \cdot \mathbf{m}_0 \mathbf{m}_1 + \varepsilon \right] \right\} \\ &\quad + \mathbf{l}_0' \cdot p_1 \cos\beta_{\max} |\tan\alpha_{\max}| \sqrt{1 + \tan^2\beta_{\max} \cdot (\mathbf{m}_0 \mathbf{m}_1)^2} \\ &\quad + \mathbf{l}_1 \cdot p_1 \cos\beta_{\max} |\sin\beta_{\max}| \left[\tan^2\alpha_{\max} - 2 |\tan\alpha_{\max} \cdot \tan\beta_{\max}| \cdot \mathbf{m}_0 \mathbf{m}_1 + \varepsilon \right] \end{aligned} \tag{10.132}$$

11 Floating Glaciers

Abstract

This chapter describes glaciers in local floating equilibrium with the general solution of balance and boundary conditions. It describes icebergs with their boundary stresses known on the entire boundary as well as the resulting properties. And it finally describes horizontally isotropic-homogeneous, infinitely extended table iceberg models with the mathematically exact, unique solutions of all relevant conditions.

How to construct the general solution of the balance and boundary conditions (2.14)–(2.16) for floating glaciers has already been presented.[1] In this chapter, some aspects of floating glaciers will be discussed using simple examples.

In the following, the z-axis is vertically upward (11.1) and the water level defines the zero level $z = 0$. The water density $\widetilde{\rho}(z)$ is horizontally homogeneous, a smooth function $\overline{\widetilde{\rho}}(z)$ below the zero level and also defined above the water level, where it vanishes by definition, caused by the jump function $\theta(-z)$ also vanishing there. (11.2) The hydrostatic water pressure $\widetilde{p}(z)$ (11.3) is also a horizontally homogeneous function, vanishing above the water level. The hydrostatic stress tensor $\widetilde{\mathbf{S}}$ (11.5) is isotropic.

—

$$\mathbf{g} = -g \cdot \mathbf{e}_z \tag{11.1}$$

[1] See Sect. 9.2.1.

P. Halfar, *Stresses in glaciers*, https://doi.org/10.1007/978-3-662-66024-9_11

$$\widetilde{\rho}(z) = \theta(-z) \cdot \overline{\widetilde{\rho}}(z) \tag{11.2}$$

$$\widetilde{p}(z) = -g \cdot \partial_z^{-1} \widetilde{\rho} = g \cdot \int_0^\infty dz' \cdot \widetilde{\rho}(z + z') \tag{11.3}$$

$$\widetilde{\rho} = 0, \widetilde{p} = 0; z \overset{\text{prec.}}{>} 0 \tag{11.4}$$

$$\widetilde{\mathbf{S}} = -\widetilde{p} \cdot \mathbf{1} \tag{11.5}$$

11.1 Glaciers in Local Floating Equilibrium

The following model of a partially floating glacier has a relatively simple general solution of the balance and boundary conditions.

In this model, the boundary surface Σ of the glacier area Ω under consideration , on which the boundary stresses are known, is simply connected[2] and consists of three parts: a top surface Σ_0, a bottom surface Σ_1 and a vertical sidewall $\Sigma_\perp$. (11.6) The top surface Σ_0 defined by a function $z_0(x, y)$ is part of the free glacier surface and lies above zero level. (11.7) The bottom surface Σ_1 defined $z_1(x, y)$ by a function $z_1(x, y)$ lies vertically below a portion of the top surface Σ_0 and below zero level in the water (11.8). The vertical sidewall $\Sigma_\perp$ extends vertically up and down from the water level. There is local buoyant equilibrium everywhere, so that on the underside Σ_1 the gravity pressure p (11.10) of the ice is equal to the water pressure $\widetilde{p}$ (11.3) and (11.11) This means that any vertical column of ice above the underside Σ_1 is supported by the water pressure .

The diagonal stress tensor field $\mathbf{S}_{\text{spez}}$ (11.17) with the negative water pressure $\widetilde{p}$ (11.3) at the first two diagonal locations and the negative gravity pressure p (11.10) of the ice at the remaining location satisfies the balance conditions (11.12). Similarly, this diagonal stress tensor field satisfies the boundary conditions on the surfaces Σ_0, Σ_1 and $\Sigma_\perp$, (11.13)–(11.15) which imply that on the boundary surface Σ the boundary stresses under water are determined by the water pressure and vanish on the free surface. The general solution $\mathbf{S}$ of the balance and boundary conditions (11.12)–(11.15) is obtained from this particular solution by adding (11.16) weightless stress tensor fields $\mathbf{T}_0$ with boundary stresses vanishing on the boundary surface Σ. $\mathbf{T}_0$ result as second derivatives of symmetric stress functions $\mathbf{A}_0$, which together with their first derivatives vanish on the boundary surface Σ. (11.18) and (11.19).

Thus, for the general solution of $\mathbf{S}$ the balance and boundary conditions (11.12)–(11.15), there is a simple representation (11.16) consisting of the special solution $\mathbf{S}_{\text{spez}}$ (11.17) and the weightless stress tensor fields $\mathbf{T}_0$. All weightless stress tensor fields $\mathbf{T}_0$ (11.18), which

[2] Therefore, no free parameters occur. See Sect. 8.1.

arise from symmetric $\mathbf{A}_0$-matrix fields (11.19),[3] lead to solutions $\mathbf{S}$ (11.16) of the balance and boundary conditions and any solution can be written in this form.[4]

—

$$\Sigma = \Sigma_0 \cup \Sigma_1 \cup \Sigma_\perp \tag{11.6}$$

$$z = z_0(x, y) > 0; \ \ \mathbf{r} \overset{\text{prec.}}{\in} \Sigma_0 \tag{11.7}$$

$$z = z_1(x, y) < 0 < z_0(x, y); \mathbf{r} \overset{\text{prec.}}{\in} \Sigma_1 \tag{11.8}$$

$$\rho = \theta(z_0 - z) \cdot \overline{\rho} \tag{11.9}$$

$$p(x, y, z) = -g \cdot \partial_z^{-1} \rho \tag{11.10}$$

$$p|_{z=z_1} \overset{\text{prec.}}{=} \tilde{p}(z_1) \Leftrightarrow \partial_z^{-1} \rho|_{z=z_1} = \partial_z^{-1} \tilde{\rho}|_{z=z_1} \tag{11.11}$$

* * *

$$\operatorname{div} \mathbf{S} = g\rho \cdot \mathbf{e}_z = -\partial_z p \cdot \mathbf{e}_z; \quad \mathbf{S} = \mathbf{S}^{\mathrm{T}} \tag{11.12}$$

$$[\mathbf{S} \cdot \mathbf{n}]_{\Sigma_0} = \mathbf{0} \tag{11.13}$$

$$[\mathbf{S} \cdot \mathbf{n}]_{\Sigma_1} = -[\tilde{p} \cdot \mathbf{n}]_{\Sigma_1} \tag{11.14}$$

$$[\mathbf{S} \cdot \mathbf{n}]_{\Sigma_\perp} = -[\tilde{p} \cdot \mathbf{n}]_{\Sigma_\perp} \tag{11.15}$$

* * *

$$\mathbf{S} = \mathbf{S}_{\text{spez}} + \mathbf{T}_0 \tag{11.16}$$

$$\mathbf{S}_{\text{spez}} = \begin{bmatrix} -\tilde{p} & 0 & 0 \\ 0 & -\tilde{p} & 0 \\ 0 & 0 & -p \end{bmatrix} \tag{11.17}$$

[3] rot $\operatorname{rot}\mathbf{A}_0$ can be calculated using formula (13.24), Chap. 13.

[4] Redundancies occur because different matrix fields $\mathbf{A}_0$ can lead to the same stress tensor field $\mathbf{T}_0$ (see Sect. 7.4).

$$\mathbf{T}_0 = \mathbf{T}_0^{\mathrm{T}} = \mathrm{rot\ rot}\mathbf{A}_0 \tag{11.18}$$

$$\mathbf{A}_0 = \mathbf{A}_0^{\mathrm{T}}; \quad \mathbf{A}_0|_\Sigma = \mathbf{0}; \quad \partial_n \mathbf{A}_0|_\Sigma = \mathbf{0} \tag{11.19}$$

11.2 Boundary Stresses on Closed Boundaries and the Global Balance Conditions for Icebergs

If the boundary stresses on the whole closed boundary $\partial\Omega$ of a glacier area Ω are known and given as boundary conditions, one has to respect, that these boundary stresses must fulfil compatibility conditions, which follow from the global balance of forces and torques. These compatibility conditions imply, that the boundary stresses on this closed boundary $\partial\Omega$ of a glacier region Ω must generate the force and torque that balance the weight and torque, respectively, of the ice mass that lies in that glacier region Ω.

The meaning of this compatibility condition shall be explained by the example of an iceberg which is in floating equilibrium, since in this case the boundary stresses on the entire closed boundary surface $\partial\Omega$ of the iceberg are known. In the corresponding compatibility conditions (11.20) and (11.21), the iceberg mass m and its center of gravity vector $\mathbf{c}$ occur.[5] Analogous conditions (11.22) and (11.23) hold for the fictitious water[6] mass with the fictitious water mass $\tilde{m}$ and with the centroid vector $\tilde{\mathbf{c}}$. Accordingly, the compatibility conditions for the boundary stresses of an iceberg in floating equilibrium imply that Archimedes' principle (11.24) holds and that the center of gravity of the iceberg and the center of gravity of the fictitious water mass are vertically superimposed (11.25).

—

$$\oint_{\partial\Omega} \tilde{\mathbf{S}}\mathbf{n} \cdot dA + m\mathbf{g} = 0 \tag{11.20}$$

$$\oint_{\partial\Omega} \mathbf{r} \times \tilde{\mathbf{S}}\mathbf{n} \cdot dA + m\mathbf{c} \times \mathbf{g} = 0 \tag{11.21}$$

[5] These are the balance conditions (2.8) and (2.9) applied to the whole iceberg, using the centroid definition (2.7). The boundary stresses $\tilde{\mathbf{S}}\mathbf{n}$ are defined by the hydrostatic tensor $\tilde{\mathbf{S}}$ (11.5), which vanishes in the half-space above the water level.

[6] The fictitious water mass fills the iceberg volume up to the height of the water level.

$$\oint_{\partial\Omega} \tilde{\mathbf{S}}\mathbf{n} \cdot dA + \tilde{m}\mathbf{g} = 0 \tag{11.22}$$

$$\oint_{\partial\Omega} \mathbf{r} \times \tilde{\mathbf{S}}\mathbf{n} \cdot dA + \tilde{m}\tilde{\mathbf{c}} \times \mathbf{g} = 0 \tag{11.23}$$

$$m\mathbf{g} = \widetilde{m}\mathbf{g} \tag{11.24}$$

$$(\mathbf{c} - \widetilde{\mathbf{c}}) \times \mathbf{g} = 0 \tag{11.25}$$

11.3 Horizontally Isotropic-Homogeneous Table Iceberg Models

In this section, we discuss horizontally isotropic-homogeneous, infinitely extended table iceberg models,[7] considering not only the balance and boundary conditions for the stresses, but also the flow law. These models lend themselves well to analysis because of their high symmetry. They are of principal theoretical importance because they have a unique solution and they are candidates for describing horizontally very extended homogeneous table icebergs at sufficient distance from their lateral margins.[8]

The models each meet the following conditions:

1. Horizontal isotropy and homogeneity of the stress tensor field
2. Balance conditions of all forces and torques (2.14) and (2.15)
3. Boundary conditions of given boundary stresses (2.16) at the surface ($z = z_0(x,y)$) and at the bottom ($z = z_1(x,y)$) of the table iceberg
4. Consideration of the lateral water pressure
5. Incompressible flow and horizontal isotropy and homogeneity of the tensor field of strain rates.
6. Flow law

[7] This means that the tensor fields of stresses and strain rates are invariant to horizontal displacements and to rotations about a vertical axis, and that the physical properties of the ice, i.e. density, temperature, etc. are horizontally homogeneous.

[8] This candidate status is based on the conjecture that the unique solution of these models arises as a limit if one assumes finite, horizontally homogeneous table icebergs and lets these table icebergs grow horizontally unrestricted. A proof of this conjecture could not be found.

11.3.1 Horizontally Isotropic-Homogeneous Stress Tensor Fields

The stress tensor fields satisfying the conditions given in points 1–3 above, (11.28) are diagonal, contain an indefinite function $S_{xx}(z)$, and contain the gravity pressure p (11.26) of the ice, which at the bottom must coincide (11.29) with the water pressure $\widetilde{p}$ (11.27). This is equivalent to Archimedes' principle. This determines the depth of immersion of the table iceberg.

—

$$p(z) = g \cdot \int_{z}^{z_0} dz' \cdot \rho(z') \tag{11.26}$$

$$\widetilde{p}(z) = g \cdot \int_{z}^{0} dz' \cdot \widetilde{\rho}(z') \tag{11.27}$$

$$\mathbf{S}(z) = \begin{bmatrix} S_{xx}(z) & 0 & 0 \\ 0 & S_{xx}(z) & 0 \\ 0 & 0 & -p(z) \end{bmatrix} \tag{11.28}$$

$$p(z_1) = \widetilde{p}(z_1) \tag{11.29}$$

11.3.2 Influence of Lateral Water Pressure

In this section, the influence of the lateral water pressure is discussed and thus the condition mentioned in point 4 above is explained.

In order to develop an idea of the influence of the lateral water pressure, one assumes that the horizontally unbounded model table iceberg and its stress tensor field emerge as a unique limiting case. One assumes finite, horizontally homogeneous[9] model table icebergs with vertical sidewalls and lets them horizontally grow unrestrictedly. In doing so, one investigates the forces and torques on oriented, vertical surfaces. Such a surface is also called an oriented curtain surface, since it resembles a curtain that is suspended from the free surface of the table iceberg and that extends to its oriented, lower seam at the base.[10]

[9]These finite model table icebergs are supposed to have horizontally homogeneous physical parameters such as ice density, temperature, etc. However, the stress tensor fields in these finite table icebergs are not horizontally homogeneous because of the influences coming from the lateral edge.

[10]The oriented surface normal **n** on such a curtain surface is defined by its oriented bottom seam. This surface normal **n** is the vector product of the oriented, horizontal, tangential unit vector at the lower seam and of the unit vector $\mathbf{e}_z$ pointing vertically upwards.

In an unbounded model table iceberg, the force $\mathbf{F}_{\mathbf{a},\,\mathbf{b}}$ (11.30) and torque $\mathbf{G}_{\mathbf{a},\,\mathbf{b}}$ (11.31) generated by the stress tensor field $\mathbf{S}$ (11.28) on an oriented curtain surface $\Gamma_{\mathbf{a},\,\mathbf{b}}$ depend only on the starting point **a** and the ending point **b** of its lower seam, but not on the remaining course of the curtain surface.[11] This independence from the course of the curtain surface exists because the force and the torque on an oriented, closed curtain surface[12] disappear, because the weight and the weight-related torque of the cylindrical ice mass, which lies within the closed curtain surface, are balanced solely by the force and the torque at the bottom.

For comparison, a finite table iceberg in floating equilibrium with vertical sidewall and horizontally homogeneous ice density is considered. In this table iceberg, the force and torque on an oriented curtain surface also depend only on the starting point **a** and end point of **b** its lower seam.[13] Although the stress tensor field in the finite table iceberg is unknown, the force and torque can be given on an oriented curtain surface that completely cuts through the table iceberg and whose lower seam thus begins at an initial point at **a** the bottom of the vertical sidewall and extends along the bottom to an end point **b** that is again at the bottom of the vertical sidewall. Since these quantities depend only on the starting point **a** and the ending point **b** of the lower seam, they can be calculated by running the lower seam along the vertical sidewall of the table iceberg from **a** to **b**. The curtain surface above this seam then lies on the vertical edge of the table iceberg, where the stresses are defined by the water pressure. Therefore, the force and torque on an oriented curtain surface completely cutting through the finite table iceberg are the same as the corresponding quantities $\tilde{\mathbf{F}}_{\mathbf{a},\mathbf{b}}$ (11.32) and $\tilde{\mathbf{G}}_{\mathbf{a},\mathbf{b}}$ (11.33), respectively, in the water body.

In the following it is assumed that the stress tensor field in such a finite model table iceberg, whose horizontal extension becomes larger and larger, converges against that stress tensor field $\mathbf{S}$ (11.28) and (11.29) in an unbounded model table iceberg, which generates the same forces (11.34) and torques (11.35) on curtain surfaces as the water pressure.[14] Thus, the longitudinal stress S_{xx} of the stress tensor field $\mathbf{S}$ (11.28) in the unbounded model table iceberg must satisfy two corresponding conditions (11.36) and (11.37).

The influence of the lateral water pressure is thus taken into account by the conditions (11.36) and (11.37) for the longitudinal stress S_{xx}, which can be interpreted as boundary conditions on the infinitely distant lateral boundary of the model table iceberg. Since in the following also the flow law plays a role, in which the deviatoric longitudinal stress σ

[11] The vector $(\mathbf{a} + \mathbf{b})/2$ in formula (11.31) leads from the origin of coordinates to the midpoint between points **a** and **b**, the vector $\mathbf{b} - \mathbf{a}$ leads from point **a** to point **b**.

[12] For a closed curtain surface, the start and end points of its bottom seam coincide.

[13] The rationale is the same as for the unlimited model table iceberg.

[14] This assumption could not be proven and therefore remains merely a plausible hypothesis.

(11.39) occurs, the conditions for the longitudinal stress S_{xx} are transformed into conditions (11.40) and (11.41) for the deviatoric longitudinal stress σ.[15]

If, in the table iceberg of thickness h (11.42) and immersion depth $\tilde{h}$ (11.43), instead of the vertical coordinate z, one introduces the dimensionless vertical coordinate λ (11.44), which has the value zero at the bottom and one at the surface, the conditions for the deviatoric longitudinal stress take a corresponding form (11.46) and (11.47).[16]

—

$$\mathbf{F_{a,b}} = \int_{\Gamma_{a,b}} \mathbf{Sn} \cdot dA = -\mathbf{e}_z \times (\mathbf{b} - \mathbf{a}) \cdot \int_{z_1}^{z_0} dz \cdot S_{xx}(z) \tag{11.30}$$

$$\mathbf{G_{a,b}} = \int_{\Gamma_{a,b}} \mathbf{r} \times \mathbf{Sn} \cdot dA = \frac{1}{2} \cdot (\mathbf{a} + \mathbf{b}) \times \mathbf{F_{a,b}} + (\mathbf{b} - \mathbf{a}) \cdot \int_{z_1}^{z_0} dz \cdot (z - z_1) \cdot S_{xx}(z) \tag{11.31}$$

$$\tilde{\mathbf{F}}_{\mathbf{a,b}} = \int_{\Gamma_{a,b}} \tilde{\mathbf{S}}\mathbf{n} \cdot dA = \mathbf{e}_z \times (\mathbf{b} - \mathbf{a}) \cdot \int_{z_1}^{0} dz \cdot \tilde{p}(z) \tag{11.32}$$

$$\tilde{\mathbf{G}}_{\mathbf{a,b}} = \int_{\Gamma_{a,b}} \mathbf{r} \times \tilde{\mathbf{S}}\mathbf{n} \cdot dA = \frac{1}{2} \cdot (\mathbf{a} + \mathbf{b}) \times \tilde{\mathbf{F}}_{\mathbf{a,b}} - (\mathbf{b} - \mathbf{a}) \cdot \int_{z_1}^{0} dz \cdot (z - z_1) \cdot \tilde{p}(z) \tag{11.33}$$

* * *

$$\mathbf{F_{a,b}} = \tilde{\mathbf{F}}_{\mathbf{a,b}} \tag{11.34}$$

$$\mathbf{G_{a,b}} = \tilde{\mathbf{G}}_{\mathbf{a,b}} \tag{11.35}$$

$$\int_{z_1}^{z_0} dz \cdot S_{xx}(z) = -\int_{z_1}^{0} dz \cdot \tilde{p}(z) \tag{11.36}$$

[15] In formulae (11.40) and (11.41) the gravity pressure $\tilde{p}(z)$ is also defined above the water level by the value zero.

[16] For a clear symbolization of the functional dependencies on the vertical variables z and λ, the same function symbols are used. For example, the vertical variation of the gravity pressure p is denoted by $p(\lambda)$ as well as by $p(z)$, with the function symbol $p(\cdot)$ having different meanings in each case. (To obtain a uniform meaning, one would have to use the more cumbersome symbol $p\left(h \cdot \lambda - \tilde{h}\right)$ instead of the simple symbol $p(\lambda)$ because of $z = h \cdot \lambda - \tilde{h}$).

$$\int_{z_1}^{z_0} \mathrm{d}z \cdot z \cdot S_{xx}(z) = -\int_{z_1}^{0} \mathrm{d}z \cdot z \cdot \widetilde{p}(z) \tag{11.37}$$

* * *

$$\mathbf{S}'(z) = \sigma(z) \cdot \begin{bmatrix} 1 & 0 & 0 \\ 0 & 1 & 0 \\ 0 & 0 & -2 \end{bmatrix} \tag{11.38}$$

$$\sigma = \frac{1}{3} \cdot (S_{xx} + p); \quad S_{xx} = 3\sigma - p \tag{11.39}$$

* * *

$$\int_{z_1}^{z_0} \mathrm{d}z \cdot \sigma(z) = \int_{z_1}^{z_0} \mathrm{d}z \cdot \frac{[p(z) - \widetilde{p}(z)]}{3} \tag{11.40}$$

$$\int_{z_1}^{z_0} dz \cdot z \cdot \sigma(z) = \int_{z_1}^{z_0} dz \cdot z \cdot \frac{[p(z) - \widetilde{p}(z)]}{3} \tag{11.41}$$

* * *

$$h \stackrel{\text{def.}}{=} z_0 - z_1 \tag{11.42}$$

$$\tilde{h} \stackrel{\text{def.}}{=} -z_1 \tag{11.43}$$

$$\lambda \stackrel{\text{def.}}{=} \frac{z + \tilde{h}}{h} \tag{11.44}$$

$$z = h \cdot \lambda - \widetilde{h} \tag{11.45}$$

* * *

$$\int_{0}^{1} \mathrm{d}\lambda \cdot \sigma(\lambda) = \int_{0}^{1} \mathrm{d}\lambda \cdot \frac{[p(\lambda) - \tilde{p}(\lambda)]}{3} \stackrel{\text{def.}}{=} C_1 \tag{11.46}$$

$$\int_{0}^{1} \mathrm{d}\lambda \cdot \lambda \cdot \sigma(\lambda) = \int_{0}^{1} \mathrm{d}\lambda \cdot \lambda \cdot \frac{[p(\lambda) - \tilde{p}(\lambda)]}{3} \stackrel{\text{def.}}{=} C_2 \tag{11.47}$$

11.3.3 Flow Velocities and Strain Rates

In this section, the horizontally isotropic-homogeneous tensor fields of the strain rates of incompressible flows are presented. These fulfil the conditions mentioned above under number 5 (Sect. 11.3).

Each vector field **v** of flow velocities defines its tensor field **D** (11.48) of strain rates. This tensor field **D** is symmetric (11.49) and it satisfies an integrability condition (11.50). This symmetry and this integrability condition are characteristic of all tensor fields **D** of strain rates, since they are not only necessary but also sufficient [2, p. 40] for there to be a velocity field **v** whose strain rates are given by **D** (11.48).[17]

The horizontally isotropic-homogeneous strain rate tensor fields **D** of incompessible flows are invariant to horizontal displacements and to rotations about a vertical axis and its trace vanishes. Thus they are defined by the horizontal strain rate d (11.51) and depend only on the dimensionless vertical coordinate λ. If one still takes into account the above mentioned integrability condition (11.50), then the horizontal strain rate d must be a linear function of the dimensionless vertical coordinate λ (11.52). This linear function d can be characterized either by its zero λ_* (11.53) and the difference Δ (11.54) between the horizontal strain rates at the surface and at the bottom of the table iceberg (11.55) or it is a constant d_c (11.56).

This gives the general form of a horizontally isotropic-homogeneous strain rate tensor field **D** of an incompressible flow (11.51), (11.55), and (11.56).

—

$$\mathbf{D} = \frac{1}{2} \cdot \left[\text{grad}\,\mathbf{v} + (\text{grad}\,\mathbf{v})^{\mathrm{T}}\right]; \quad D_{ik} = \frac{1}{2} \cdot (\partial_k v_i + \partial_i v_k) \tag{11.48}$$

$$\mathbf{D} = \mathbf{D}^{\mathrm{T}} \tag{11.49}$$

$$\text{rot rot}\,\mathbf{D} = 0 \tag{11.50}$$

* * *

$$\mathbf{D}(\lambda) = d(\lambda) \cdot \begin{bmatrix} 1 & 0 & 0 \\ 0 & 1 & 0 \\ 0 & 0 & -2 \end{bmatrix} \tag{11.51}$$

$$\partial_\lambda^2 d(\lambda) = 0 \tag{11.52}$$

[17] This compatibility theorem also follows from Sect. 14.1, with $\mathbf{D} = \mathbf{A}_+^{\bullet}$.

* * *

$$d(\lambda_*) \overset{\text{def.}}{=} 0 \tag{11.53}$$

$$\Delta \overset{\text{def.}}{=} d|_{\lambda=1} - d|_{\lambda=0} \tag{11.54}$$

$$d(\lambda) = \Delta \cdot (\lambda - \lambda_*); \quad \Delta \overset{\text{prec.}}{\neq} 0 \tag{11.55}$$

$$d(\lambda) = d_c; \quad \Delta \overset{\text{prec.}}{=} 0 \tag{11.56}$$

11.3.4 The Unique Solution, Even with Generalized Flow Law and with Generalized Lateral Boundary Conditions

In this section it is shown that there is a unique solution to the model conditions (see Sect. 11.3). This unique solvability results from the general mathematical structure of the model, and is thus proved for a generalised model, where the generalisation extends to the flow law and to the values of the parameters which take into account the influence of the lateral water pressure. (Another, simple and thus elegant proof of uniqueness is given in Sect. 20.2). The fact that this generalized model also has a unique solution is not only of theoretical interest, but may also play a role if, for whatever reason, one should be forced to use a flow law other than the usual one. In this case, too, a unique solution is guaranteed, provided that the flow law remains within the framework described below.

The generalization of the lateral boundary conditions, (11.46) and (11.47) which consider the influence of the lateral water pressure, concerns the thereby occurring constants C_1 and C_2 which are defined by the gravity pressures of ice and water. In the generalized lateral boundary conditions (11.57) and (11.58) these constants C_1 and C_2 are independent and can take arbitrary values. These constants C_1 and C_2 define the forces (11.30) and torques (11.31) acting on curtain surfaces in the table iceberg.[18]

The generalization of the flow law takes into account that the deviatoric stress tensor $\mathbf{S}'$ (11.38) is defined by the deviatoric horizontal stress σ and that the strain rate tensor $\mathbf{D}$ (11.51) is defined by the horizontal strain rate d. So, in this case, any flow law that is possible in principle must specify the deviatoric horizontal stress σ as a function Φ (11.61) of the horizontal strain rate d. This function Φ can be different on different horizontal planes,[19] so it can still depend explicitly on the dimensionless vertical coordinate λ (11.44).

---

[18] See Sect. 11.3.2. the constants C_1 and C_2 define the integrals (11.59) and (11.60) on the longitudinal stress and hence the forces (11.30) and torques (11.31) on curtain surfaces.

[19] The concrete flow law depends on the ice temperature, which may be different at different horizontal planes. Therefore, the generalized flow law can depend on the dimensionless vertical coordinate λ explicitly.

The generalization of the flow law is that this function Φ is arbitrary except for the following conditions[20]:

- When the horizontal strain rate d vanishes, the deviatoric horizontal stress σ also vanishes (11.62).
- As the horizontal strain rate d increases, the deviatoric horizontal stress σ also increases (11.63).
- If the horizontal strain rate d grows or falls unrestrictedly, the deviatoric horizontal stress σ also grows or falls unrestrictedly, and uniformly for all λ.

This means that there is a function $\Phi(d)$ of the horizontal strain rate d, independent of λ, whose magnitude tends to infinity, (11.65) if d tends to infinity,[21] where this magnitude of $\Phi(d)$ should not be greater than the magnitude of the function $\Phi(\lambda, d)$ (11.64).

The horizontal strain rate d is a linear function (11.55) and (11.56) of the dimensionless vertical coordinate λ. Uniqueness of solution means that there is exactly one linear function $d(\lambda)$ such that the deviatoric horizontal stress σ (11.61) calculated from it using the generalized flow law satisfies the generalized lateral boundary conditions (11.57) and (11.58).

If the horizontal strain rate d is spatially constant (11.71), then the generalized lateral boundary conditions (11.57) and (11.58) associated with the generalized flow law (11.61) take forms (11.72) and (11.73) that can be satisfied only if there exists a relation (11.75) between the constants C_1 and C_2 defined by a function χ (11.70).[22] Thus, if the constants C_1 and C_2 satisfy this relation, there is a unique spatially constant longitudinal strain rate d_c (11.74) as a solution of the generalized lateral boundary conditions associated with the generalized flow law. This solution is unique since there is no non-constant longitudinal strain rate as a solution. The latter follows from the following investigation of the spatially non-constant longitudinal strain rates.

If the horizontal strain rate d is not spatially constant (11.76), the generalized lateral boundary conditions (11.57) and (11.58) in conjunction with the generalized flow law (11.61) take forms (11.77) and (11.78) in which the functions $I_1(\lambda_*, \Delta)$ (11.66) and $I_2(\lambda_*, \Delta)$ (11.67) should take the values C_1 and C_2, respectively.[23] The solution of these conditions can be represented graphically (cp. Fig. 11.2). There is therefore a unique solution of these

[20] These conditions are not further substantiated but justified by the fact that the flow law of the ice (see Sect. 11.3.5.) fulfils these conditions.

[21] By definition, a sequence of numbers tends to infinity if the sequence of its reciprocals converges to zero.

[22] The function χ is defined by the flow law function Φ. This function χ is a nested function and consists of the inverse function of a function K_1 and of a function K_2. These functions K_1, K_2, and χ are monotonically increasing, vanishing for vanishing argument, and their range of values extends from minus infinity to plus infinity. This follows from properties (11.62)–(11.64) of function Φ .

[23] Functions I_1 and I_1 are discussed in Sect. 20.1.

conditions[24] because the function values $I_2(\lambda_*, \Delta)$ vary monotonically from minus infinity to plus infinity as one traverses the I_1 level lines in the λ_*-Δ-coordinate system at constant I_1 values C_1 (cp. Fig. 11.1.) from bottom to top. In each case the special I_2 value C_2, (11.75) to which there is a spatially constant solution, is not reached, but is aimed at as a limit value when one approaches the abscissa asymptotically on the I_1 level lines. So also in this special case (11.75) – as already mentioned above – there is only one solution, namely the spatially constant solution.

—

$$\int_0^1 d\lambda \cdot \sigma = C_1 \tag{11.57}$$

$$\int_0^1 d\lambda \cdot \lambda \cdot \sigma = C_2 \tag{11.58}$$

$$\int_{z_1}^{z_0} dz \cdot S_{xx} = 3h \cdot C_1 - h \cdot \int_0^1 d\lambda \cdot p \tag{11.59}$$

$$\int_{z_1}^{z_0} \mathrm{d}z \cdot (z - z_1) \cdot S_{xx} = 3h^2 \cdot C_2 - h^2 \cdot \int_0^1 \mathrm{d}\lambda \cdot \lambda \cdot p \tag{11.60}$$

* * *

$$\sigma = \Phi(\lambda, d) \tag{11.61}$$

$$\Phi(\lambda, 0) \overset{\text{prec.}}{=} 0 \tag{11.62}$$

$$\frac{\Phi(\lambda, d') - \Phi(\lambda, d)}{d' - d} \overset{\text{prec.}}{>} 0; \quad d' \neq d \tag{11.63}$$

$$| \breve{\Phi}(d) | \overset{\text{prec.}}{\leq} | \Phi(\lambda, d) | \tag{11.64}$$

$$d \to \pm\infty : \quad | \breve{\Phi}(d) | \overset{\text{prec.}}{\to} \infty \tag{11.65}$$

* * *

[24] See the detailed justification in Sect. 20.2.

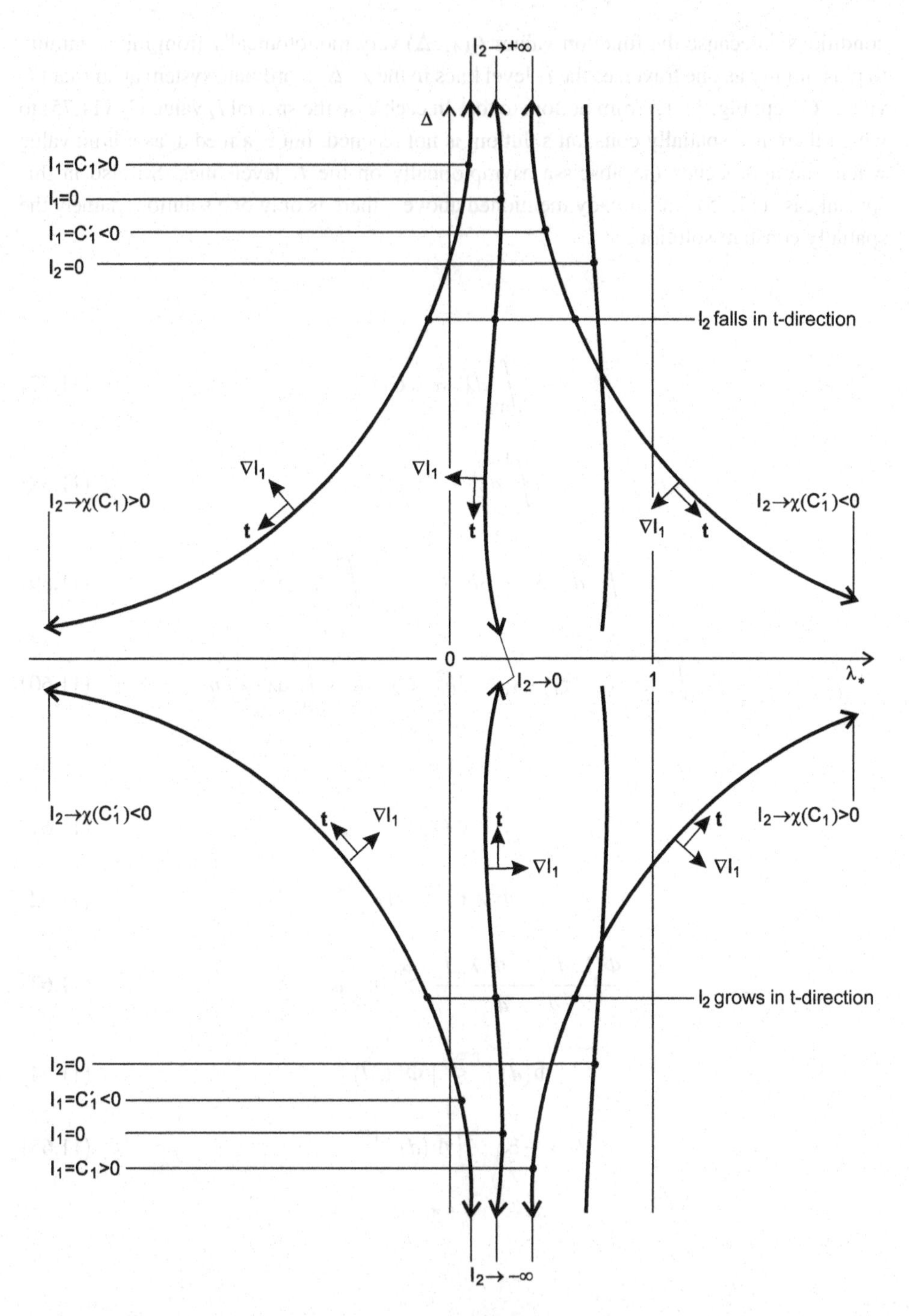

Fig. 11.1 The λ_*-Δ-coordinate system with level lines of the function $I_1(\lambda_*, \Delta)$, with the zero level line of the function $I_2(\lambda_*, \Delta)$, with the gradient and tangent vectors of the I_1 level lines and with information on the course of the function I_2 on the I_1 level lines (see Sect. 20.1)

$$I_1(\lambda_*, \Delta) \stackrel{\text{def}}{=} \int_0^1 \mathrm{d}\lambda \cdot \Phi[\lambda, \Delta(\lambda - \lambda_*)] \tag{11.66}$$

$$I_2(\lambda_*, \Delta) \stackrel{\text{def}}{=} \int_0^1 \mathrm{d}\lambda \cdot \lambda \cdot \Phi[\lambda, \Delta(\lambda - \lambda_*)] \tag{11.67}$$

$$K_1(d) \stackrel{\text{def}}{=} \int_0^1 \mathrm{d}\lambda \cdot \Phi[\lambda, d] \tag{11.68}$$

$$K_2(d) \stackrel{\text{def}}{=} \int_0^1 \mathrm{d}\lambda \cdot \lambda \cdot \Phi[\lambda, d] \tag{11.69}$$

$$\chi(C_1) \stackrel{\text{def.}}{=} K_2\left[\overset{-1}{K_1}(C_1)\right] \tag{11.70}$$

* * *

$$\Delta = 0: \quad d(\lambda) \stackrel{(11.56)}{=} d_c \tag{11.71}$$

$$K_1(d_c) = C_1 \tag{11.72}$$

$$K_2(d_c) = C_2 \tag{11.73}$$

$$d_c = \overset{-1}{K_1}(C_1) \tag{11.74}$$

$$C_2 \stackrel{(11.70)}{=} \chi(C_1) \tag{11.75}$$

* * *

$$\Delta \neq 0: \qquad d(\lambda) \stackrel{(11.55)}{=} \Delta \cdot (\lambda - \lambda_*) \tag{11.76}$$

$$I_1(\lambda_*, \Delta) = C_1 \tag{11.77}$$

$$I_2(\lambda_*, \Delta) = C_2 \tag{11.78}$$

11.3.5 The Flow Law

The flow law[25] [4, pp. 91, 96] describes a relation (11.82) between the deviatoric stress tensor $\mathbf{S}'$ (11.79) and the tensor $\mathbf{D}$ (11.48) of strain rates, where the invariants S' (11.80) and D (11.81) of these tensors and the flow law parameter A occur. Corresponding relations hold between the invariants D and S' (11.83).

In the horizontally isotropic case, the flow law defines the horizontal component σ of the deviatoric stress tensor field $\mathbf{S}'$ (11.85) as a function (11.88) of the horizontal component d of the tensor field $\mathbf{D}$ (11.84) of the strain rates.[26] For the functions that play a role according to the general investigation carried out above,[27] special expressions are obtained (11.89)–(11.95).

—

$$\mathbf{S}' \overset{\text{def.}}{=} \mathbf{S} - \frac{1}{3} \cdot \text{trace}\,(\mathbf{S}) \cdot \mathbf{1} \tag{11.79}$$

$$S' \overset{\text{def.}}{=} \frac{1}{\sqrt{2}} \cdot \sqrt{S'_{ik} S'_{ik}} = \frac{1}{\sqrt{2}} \cdot \sqrt{\text{trace}\,\left(\mathbf{S}'^T \mathbf{S}'\right)} \tag{11.80}$$

$$D \overset{\text{def.}}{=} \frac{1}{\sqrt{2}} \cdot \sqrt{D_{ik} D_{ik}} = \frac{1}{\sqrt{2}} \cdot \sqrt{\text{trace}\,\left(\mathbf{D}^T \mathbf{D}\right)} \tag{11.81}$$

$$\mathbf{D} = A \cdot S'^{n-1} \cdot \mathbf{S}'; \quad \mathbf{S}' = A^{-1/n} \cdot D^{-1+1/n} \cdot \mathbf{D} \tag{11.82}$$

$$D = A \cdot S'^n; \quad S' = A^{-1/n} \cdot D^{1/n} \tag{11.83}$$

* * *

$$\mathbf{D} = d \cdot \begin{bmatrix} 1 & 0 & 0 \\ 0 & 1 & 0 \\ 0 & 0 & -2 \end{bmatrix} \tag{11.84}$$

$$\mathbf{S}' = \sigma \cdot \begin{bmatrix} 1 & 0 & 0 \\ 0 & 1 & 0 \\ 0 & 0 & -2 \end{bmatrix} \tag{11.85}$$

$$S' = \sqrt{3} \cdot \mid \sigma \mid \tag{11.86}$$

---

[25] A general discussion of possible flow laws is given by Serrin [5, pp. 230–236].

[26] The parameter A depends on the temperature and thus on the dimensionless vertical coordinate λ.

[27] See Sect. 11.3.4.

$$D = \sqrt{3} \cdot |\, d \,| \tag{11.87}$$

$$\sigma = \sqrt{3}^{(-1+1/n)} \cdot A^{-1/n}(\lambda) \cdot |d|^{1/n} \cdot \operatorname{sign}(d) \tag{11.88}$$

* * *

$$\Phi(\lambda, d) = \sqrt{3}^{(-1+1/n)} \cdot A^{-1/n}(\lambda) \cdot |d|^{1/n} \cdot \operatorname{sign}(d) \tag{11.89}$$

$$I_1(\lambda_*, \Delta) = \sqrt{3}^{(-1+1/n)} \cdot |\Delta|^{1/n} \cdot \operatorname{sign}(\Delta) \cdot \int_0^1 d\lambda \cdot A^{-1/n}(\lambda) \cdot |\lambda - \lambda_*|^{1/n} \cdot \operatorname{sign}(\lambda - \lambda_*) \tag{11.90}$$

$$I_2(\lambda_*, \Delta) = \sqrt{3}^{(-1+1/n)} \cdot |\Delta|^{1/n} \cdot \operatorname{sign}(\Delta) \cdot \int_0^1 d\lambda \cdot \lambda \cdot A^{-1/n}(\lambda) \cdot |\lambda - \lambda_*|^{1/n} \cdot \operatorname{sign}(\lambda - \lambda_*) \tag{11.91}$$

$$K_1(d) = \sqrt{3}^{(-1+1/n)} \cdot |d|^{1/n} \cdot \operatorname{sign}(d) \cdot \int_0^1 d\lambda \cdot A^{-1/n}(\lambda) \tag{11.92}$$

$$\overset{-1}{K_1}(C_1) = \sqrt{3}^{(n-1)} \cdot \left[\int_0^1 d\lambda \cdot A^{-1/n}(\lambda)\right]^{-n} \cdot |C_1|^n \cdot \operatorname{sign}(C_1) \tag{11.93}$$

$$K_2(d) = \sqrt{3}^{(-1+1/n)} \cdot |d|^{1/n} \cdot \operatorname{sign}(d) \cdot \int_0^1 d\lambda \cdot \lambda \cdot A^{-1/n}(\lambda) \tag{11.94}$$

$$\chi(C_1) = \frac{\int_0^1 d\lambda \cdot \lambda \cdot A^{-1/n}(\lambda)}{\int_0^1 d\lambda \cdot A^{-1/n}(\lambda)} \cdot C_1 \tag{11.95}$$

11.3.6 Calculation of the Solution

If one considers the flow law, then the parameters λ_* and Δ of the spatially non-constant horizontal strain rate d (11.98) are determined by corresponding governing equations (11.100) and (11.101). For this purpose, the vertical course $A(\lambda)$ of the flow law parameter and the two constants C_1 and C_2 must be known. In the special case, if the two constants C_1 and C_2 satisfy a corresponding relation (11.102), a spatially constant, horizontal strain rate d_c is established, which is also determined by a equation (11.105) following from the flow law.

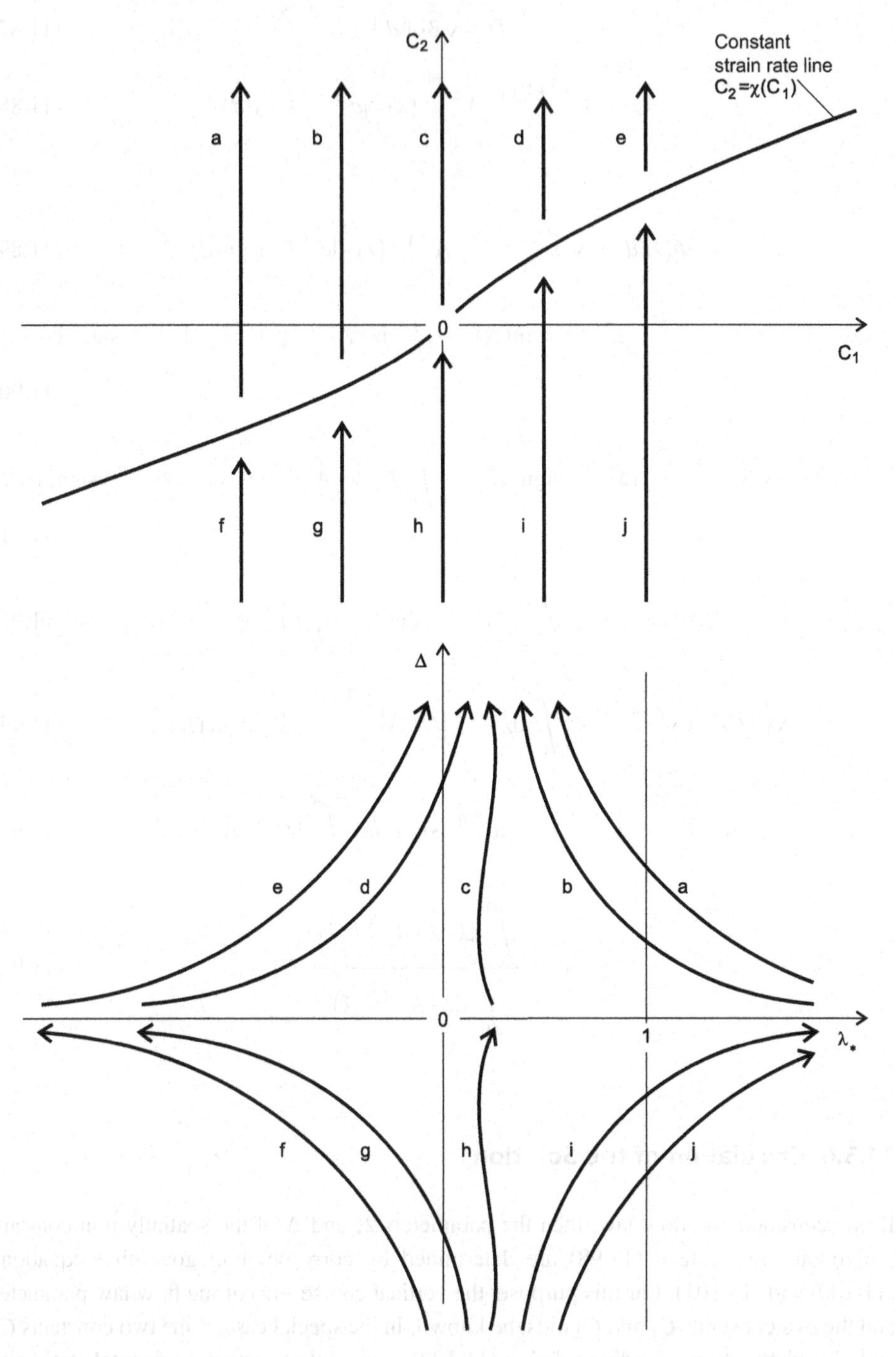

Fig. 11.2 Graphical representation of the solution of the lateral boundary conditions. To each costant pair or point (C_1, C_2) in the $C_1\_C_2$ coordinate system that does not lie on the line of spatially constant strain rates is assigned a solution pair or point (λ_*, Δ) in the λ_*_Δ coordinate system. With this assignment, the lines "a" to "j" running parallel to the ordinate in the $C_1\_C_2$ coordinate system merge into the corresponding level lines of the function I_1 in the λ_* $-\Delta$- coordinate system

In the following, let the densities of ice and of water be given by the spatially constant values ρ_c and $\widetilde{\rho}_c$, respectively The two constants C_1 and C_2[28] are determined by the gravity pressures of ice and water and thus by the corresponding densities (11.107) and (11.108). The horizontal strain rates d can be classified by the alternatives (11.110) to spatially constant strain rates[29]:

- The horizontal strain rate d increases upwards ($\Delta > 0$):
  d is expansive at the surface.[30] This case occurs, for example, if the parameter A of the flow law is also spatially constant due to spatially constant ice temperature.[31]
- The horizontal strain rate d is spatially constant ($\Delta = 0$):
  d (11.104) and (11.105) is expansive. This special case can only occur if the parameter A of the flow law decreases sufficiently strongly from the bottom to the surface,[32] i.e. the temperature decreases sufficiently strongly towards the top.[33]
- The horizontal strain rate d decreases towards the top ($\Delta < 0$):
  d is expansive at the bottom.[34] In this case, the parameter A of the flow law and the temperature must decrease upwards even more than in the case of a spatially constant strain rate.[35].

In each case, there is a zone where the horizontal strain rate d is expansive. This zone extends downward from the surface as the horizontal strain rate decreases downward and upward from the bottom as the horizontal strain rate decreases upward. In this case, this zone can penetrate the entire table iceberg, as in the case of spatially constant horizontal strain rates.

---

[28] See Sect. 20.4. The more general case of spatially non-constant densities is also treated there.

[29] C_1 and C_2 are positive and the sign of Δ is obtained from Fig. 11.2.

[30] In the C_1-C_2 coordinate system (s. Fig. 11.2), one is at a point to the right of the ordinate because the constant C_1 (11.107) is positive. The line of constant strain rates passes under this point. Therefore, in the λ_*-Δ-coordinate system, one is above the abscissa to the left of the line "c". Consequently, λ_* is less than 1 and Δ is positive, which is why the strain rate d (11.98) is positive at the surface ($\lambda = 1$).

[31] Then for the fraction in (11.110) we get the value 0.5, which is smaller than the value of $(1 + \rho_c/\widetilde{\rho}_c)/3 \approx 1{,}9/3 = 0{,}633$.

[32] In order to satisfy the corresponding condition (11.110), the fraction in this condition, which would have the value 1/2 if A were spatially constant, must reach the value of $(1 + \rho_c/\widetilde{\rho}_c)/3 \approx 1{,}9/3 = 0{,}63$, which is why the negative power $A^{-1/n}(\lambda)$ must increase sufficiently with increasing λ.

[33] Lower temperatures mean lower A values [4, p. 97]. The ice therefore becomes tougher towards the top.

[34] In this alternative, the line of constant strain rates runs above the relevant point in the C_1_C_2 coordinate system (see footnote 30). In the λ_*-Δ-coordinate system one is then below the abscissa to the right of the line "h". Therefore, λ_* is greater than 0 and Δ is negative, which is why the strain rate d (11.98) at the bottom ($\lambda = 0$) is positive.

[35] To satisfy the corresponding inequality (11.110), the fraction in this inequality must exceed the value of $(1 + \rho_c/\widetilde{\rho}_c)/3 \approx 1{,}9/3 = 0{,}63$, so the negative power $A^{-1/n}(\lambda)$ must increase correspondingly as λ increases.

Table 11.1 Characteristic parameters of the horizontal strain rates d in a table iceberg with linearly varying negative power $A^{-1/n}$ of the flow law parameter in vertical direction at a bottom temperature of 0 °C and at different surface temperatures T. The exponent n, the spatially constant densities ρ_c of ice and $\tilde{\rho}_c$ of water and the ice thickness h have the given values. For each surface temperature, the dimensionless vertical coordinate λ_* of the plane of vanishing strain rates, the difference Δ in horizontal strain rates d between the surface and the bed, and the horizontal strain rates d themselves at the bed and at the surface are given. (The temperature dependent values A_T *of* the flow law parameter are given by Paterson [4, p. 97] with an accuracy of two decimal places, here they are used as exact values)

$n = 3;\ \rho_c = 0.9t/m^3;\ \tilde{\rho}_c = 1\,t/m^3;\quad h = 500\ m;\ T(\text{Sole}) = 0^\circ C$

T (surface)	$A_T \cdot 10^{18}$	λ_*	Δ	d (sole)	d (surface)
°C	s^{-1} $(kPa)^{-3}$		a^{-1}	a^{-1}	a^{-1}
0	6800	0.09	$9.85 \cdot 10^{-1}$	$-8.38 \cdot 10^{-2}$	$9.02 \cdot 10^{-1}$
−2	2400	0.06	$4.33 \cdot 10^{-1}$	$-2.73 \cdot 10^{-2}$	$4.06 \cdot 10^{-1}$
−5	1600	0.05	$3.08 \cdot 10^{-1}$	$-1.60 \cdot 10^{-2}$	$2.92 \cdot 10^{-1}$
−10	490	0.01	$1.08 \cdot 10^{-1}$	$-1.30 \cdot 10^{-3}$	$1.07 \cdot 10^{-1}$
−15	290	−0.03	$6.51 \cdot 10^{-2}$	$1.63 \cdot 10^{-3}$	$6.67 \cdot 10^{-2}$
−20	170	−0.11	$3.64 \cdot 10^{-2}$	$3.86 \cdot 10^{-3}$	$4.03 \cdot 10^{-2}$
−25	94	−0.26	$1.78 \cdot 10^{-2}$	$4.71 \cdot 10^{-3}$	$2.25 \cdot 10^{-2}$
−30	51	−0.57	$7.74 \cdot 10^{-3}$	$4.42 \cdot 10^{-3}$	$1.22 \cdot 10^{-2}$
−35	27	−1.31	$2.73 \cdot 10^{-3}$	$3.58 \cdot 10^{-3}$	$6.30 \cdot 10^{-3}$
−40	14	−4.61	$5.63 \cdot 10^{-4}$	$2.59 \cdot 10^{-3}$	$3.16 \cdot 10^{-3}$
	9.3	$\pm\infty$	0	$2.05 \cdot 10^{-3}$	$2.05 \cdot 10^{-3}$
−45	7.3	9.72	$-1.81 \cdot 10^{-4}$	$1.76 \cdot 10^{-3}$	$1.58 \cdot 10^{-3}$
−50	3.6	3.12	$-3.47 \cdot 10^{-4}$	$1.08 \cdot 10^{-3}$	$7.37 \cdot 10^{-4}$

The strain rates are defined by the vertical variation of the flow law parameter A, i.e. by its dependence $A(\lambda)$ on the dimensionless vertical coordinate λ. In the following examples, the negative power $A^{-1/n}(\lambda)$ is supposed to vary linearly in the vertical direction, (11.111) where a temperature of 0 °C with the corresponding value of $A_{0^\circ C}$ of the flow law parameter A is supposed to prevail at the bottom and a temperature T with the corresponding value A_T of the flow law parameter A *is* supposed to prevail at the surface.[36] Horizontal strain rates d are calculated at different surface temperatures T (s. Table 11.1.).[37] Their dependence on the surface temperature T follows the pattern described above, where the surface tempera-

[36]With the help of this approach (11.111), only the properties of the table iceberg model are to be illustrated by means of concrete examples. The extent to which this approach corresponds to reality is not the subject of this investigation.
[37]The calculation shall be made in accordance with the procedure set out in Sect. 20.3.

ture at which spatially constant horizontal strain rates occur is between −40 °C and − 45 °C.[38] The surface temperature of −45°C leads to spatially almost constant strain rates, since in this case $|\lambda_*|$ is large compared to 1, and therefore the magnitude of the relative spatial variation of the strain rates (11.99) is small compared to 1.

The horizontal strain rates are expansive ($d > 0$) everywhere in the table iceberg, except at surface temperatures from 0 °C to −10 °C. At these surface temperatures, the horizontal strain rates, which are expansive at the surface, become compressive towards the bottom, since the plane of vanishing horizontal strain rates lies between the bottom and the surface ($0 < \lambda_* < 1$).

At the surface, the horizontal strain rate is expansive and decreases with decreasing surface temperature. At the bottom, on the other hand, the horizontal strain rate is compressive at a surface temperature of 0 °C and increases with decreasing surface temperature up to a surface temperature of −25 °C, at which it is expansive. Only at surface temperatures below −25 °C does it decrease with decreasing surface temperature.

—

$$\frac{\int_0^1 \mathrm{d}\lambda \cdot \lambda \cdot A^{-1/n}(\lambda)}{\int_0^1 \mathrm{d}\lambda \cdot A^{-1/n}(\lambda)} \cdot C_1 \overset{\text{prec.}}{\neq} C_2 : \tag{11.96}$$

$$\Delta \neq 0 \tag{11.97}$$

$$d(\lambda) = \Delta \cdot (\lambda - \lambda_*) \tag{11.98}$$

$$\frac{d|_{\lambda=1} - d|_{\lambda=0}}{d|_{\lambda=0}} = -\frac{1}{\lambda_*} \tag{11.99}$$

$$\frac{\int_0^1 \mathrm{d}\lambda \cdot \lambda \cdot A^{-1/n}(\lambda) \cdot |\lambda - \lambda_*|^{1/n} \cdot \operatorname{sign}(\lambda - \lambda_*)}{\int_0^1 \mathrm{d}\lambda \cdot A^{-1/n}(\lambda) \cdot |\lambda - \lambda_*|^{1/n} \cdot \operatorname{sign}(\lambda - \lambda_*)} = \frac{C_2}{C_1} \tag{11.100}$$

[→ (11.77), (11.78), (11.90), (11.91)]

$$|\Delta|^{1/n} \operatorname{sign}(\Delta) = C_1 \sqrt{3}^{(1-1/n)} \left[\int_0^1 \mathrm{d}\lambda \cdot A^{-1/n}(\lambda) \cdot |\lambda - \lambda_*|^{1/n} \cdot \operatorname{sign}(\lambda - \lambda_*) \right]^{-1} \tag{11.101}$$

[→(11.77), (11.90)]

[38] This is the temperature at which the flow law parameter a has the value $9,3 \cdot 10^{-18}\,\mathrm{s}^{-1}(\mathrm{kPa})^{-3}$ (see Chap. 20, Eq. (20.54)).

* * *

$$\frac{\int_0^1 \mathrm{d}\lambda \cdot \lambda \cdot A^{-1/n}(\lambda)}{\int_0^1 \mathrm{d}\lambda \cdot A^{-1/n}(\lambda)} \cdot C_1 \overset{\text{prec.}}{=} C_2 : \tag{11.102}$$

$$\Delta = 0 \tag{11.103}$$

$$d = d_c \tag{11.104}$$

$$|\mathrm{d_c}|^{1/n} \cdot \mathrm{sign(d_c)} = C_1 \sqrt{3}^{(1-1/n)} \left[\int_0^1 \mathrm{d}\lambda \cdot A^{-1/n}(\lambda) \right]^{-1} \tag{11.105}$$

[→ (11.72), (11.92)]

* * *

$$\frac{\rho_c}{\widetilde{\rho}_c} = \frac{\widetilde{h}}{h} \tag{11.106}$$

$$C_1 \overset{(20.79)}{=} \frac{1}{6} \cdot gh\rho_c \cdot \left(1 - \frac{\rho_c}{\tilde{\rho}_c}\right) > 0 \tag{11.107}$$

$$C_2 \overset{(20.80)}{=} \frac{1}{18} \cdot gh\rho_c \cdot \left(1 - \frac{\rho_c^2}{\tilde{\rho}_c^2}\right) \tag{11.108}$$

$$\frac{C_2}{C_1} = \frac{1}{3} \cdot \left(1 + \frac{\rho_c}{\widetilde{\rho}_c}\right) \tag{11.109}$$

* * *

$$C_2 \overset{>}{\underset{<}{=}} \chi(C_1) \Leftrightarrow \underbrace{\frac{1}{3}\left(1 + \frac{\rho_c}{\tilde{\rho}_c}\right)}_{\approx 1{,}9/3 = 0{,}633} \overset{>}{\underset{<}{=}} \frac{\displaystyle\int_0^1 \mathrm{d}\lambda \cdot \lambda \cdot A^{-1/n}(\lambda)}{\displaystyle\int_0^1 \mathrm{d}\lambda \cdot A^{-1/n}(\lambda)} \Leftrightarrow \Delta \overset{>}{\underset{<}{=}} 0 \tag{11.110}$$

* * *

$$A^{-1/n}(\lambda) = A_{0^\circ\mathrm{C}}^{-1/n} + \lambda \cdot \left[A_T^{-1/n} - A_{0^\circ\mathrm{C}}^{-1/n}\right] \tag{11.111}$$

Part IV

Mathematical Appendix

Vectors and Tensors

12

Abstract

This chapter contains some definitions and calculation rules of vector and tensor calculus.

In the definitions and calculation rules (12.1)–(12.28) given below, the following terms are used (See also appendix A):

- δ_{ij}: Components of the unit tensor (12.1)
- ε_{ijk}: Components of the antisymmetric tensor (12.2)
- $\mathbf{e}_1$, $\mathbf{e}_2$, $\mathbf{e}_3$: orthonormal basis (12.4)
- **H**, **K**: any tensors
- $\mathbf{H}_+$, $\mathbf{H}_-$: symmetrical or antisymmetrical part of the tensor **H** (12.12)
- **H**: vector associated with the antisymmetric part of **H** (12.11)
- **n**: vector of length 1
- **P** Projector on **n** (12.21)
- **Q**: Projector on the plane perpendicular to **n** (12.22)
- **r**: position vector (12.9)
- **u**, **v**: any vectors (12.6)
- **u**: antisymmetric tensor associated with **u** the vector **u** (12.7)

P. Halfar, *Stresses in glaciers*, https://doi.org/10.1007/978-3-662-66024-9_12

—

$$\delta_{ij} = \begin{cases} 1; & i=j \\ 0; & i \neq j \end{cases} \tag{12.1}$$

$$\varepsilon_{ijk} = \begin{cases} 1; & i, j, k \text{ even permutation of } 1, 2, 3 \\ -1; & i, j \text{ odd p.of } 1, 2, 3 \\ 0; & \text{otherwise} \end{cases} \tag{12.2}$$

$$\varepsilon_{ijk} \cdot \varepsilon_{ilm} = \delta_{jl} \cdot \delta_{km} - \delta_{jm} \cdot \delta_{kl} \tag{12.3}$$

* * *

$$\mathbf{e}_1 = \begin{bmatrix} 1 \\ 0 \\ 0 \end{bmatrix}; \mathbf{e}_2 = \begin{bmatrix} 0 \\ 1 \\ 0 \end{bmatrix}; \mathbf{e}_3 = \begin{bmatrix} 0 \\ 0 \\ 1 \end{bmatrix} \tag{12.4}$$

$$\not{\mathbf{e}}_1 \overset{\text{def}}{=} \begin{bmatrix} 0 & 0 & 0 \\ 0 & 0 & -1 \\ 0 & 1 & 0 \end{bmatrix}; \not{\mathbf{e}}_2 \overset{\text{def}}{=} \begin{bmatrix} 0 & 0 & 1 \\ 0 & 0 & 0 \\ -1 & 0 & 0 \end{bmatrix}; \not{\mathbf{e}}_3 \overset{\text{def}}{=} \begin{bmatrix} 0 & -1 & 0 \\ 1 & 0 & 0 \\ 0 & 0 & 0 \end{bmatrix} \tag{12.5}$$

$$\mathbf{u} = \begin{bmatrix} u_1 \\ u_2 \\ u_3 \end{bmatrix} = u_i \cdot \mathbf{e}_i \tag{12.6}$$

$$\not{\mathbf{u}} \overset{\text{def}}{=} \begin{bmatrix} 0 & -u_3 & u_2 \\ u_3 & 0 & -u_1 \\ -u_2 & u_1 & 0 \end{bmatrix} = u_i \cdot \not{\mathbf{e}}_i; \; \not{u}_{ij} = -\epsilon_{ijk} \cdot u_k \tag{12.7}$$

$$\mathbf{u} = -\frac{1}{2} \begin{bmatrix} \not{u}_{23} - \not{u}_{32} \\ \not{u}_{31} - \not{u}_{13} \\ \not{u}_{12} - \not{u}_{21} \end{bmatrix} = - \begin{bmatrix} \not{u}_{23} \\ \not{u}_{31} \\ \not{u}_{12} \end{bmatrix}; \; u_i = -\frac{1}{2} \epsilon_{ijk} \cdot \not{u}_{jk} \tag{12.8}$$

$$\mathbf{r} = \begin{bmatrix} x_1 \\ x_2 \\ x_3 \end{bmatrix} = x_i \cdot \mathbf{e}_i \tag{12.9}$$

$$\not{\mathbf{r}} = \begin{bmatrix} 0 & -x_3 & x_2 \\ x_3 & 0 & -x_1 \\ -x_2 & x_1 & 0 \end{bmatrix} = x_i \cdot \not{\mathbf{e}}_i; \; \not{\mathbf{r}}_{ij} = -\epsilon_{ijk} \cdot x_k \tag{12.10}$$

$$\not{\mathbf{H}} \overset{\text{def}}{=} -\frac{1}{2}\begin{bmatrix} H_{23}-H_{32} \\ H_{31}-H_{13} \\ H_{12}-H_{21} \end{bmatrix}; \not{\mathbf{H}}_i = -\frac{1}{2}\epsilon_{ijk}\cdot H_{jk} \tag{12.11}$$

$$\mathbf{H}_+ \overset{\text{def.}}{=} \frac{1}{2}\left(\mathbf{H}+\mathbf{H}^T\right); \mathbf{H}_- \overset{\text{def.}}{=} \frac{1}{2}\left(\mathbf{H}-\mathbf{H}^{\mathrm{T}}\right) \tag{12.12}$$

* * *

$$\not{\mathbf{u}}\cdot\mathbf{v} = \mathbf{u}\times\mathbf{v} \tag{12.13}$$

$$\not{\mathbf{u}}\cdot\not{\mathbf{v}} = \mathbf{v}\mathbf{u}^{\mathrm{T}} - (\mathbf{u}\mathbf{v})1 \tag{12.14}$$

$$\not{\mathbf{w}} = \mathbf{v}\mathbf{u}^{\mathrm{T}} - \mathbf{u}\mathbf{v}^{\mathrm{T}}; \mathbf{w} \overset{\text{prec.}}{=} \mathbf{u}\times\mathbf{v} \tag{12.15}$$

$$\text{trace }\mathbf{H} = \text{trace }\mathbf{H}^{\mathrm{T}} = \text{trace }\mathbf{H}_+ \tag{12.16}$$

$$\text{trace }\mathbf{HK} = \text{trace }\mathbf{KH} \tag{12.17}$$

$$\text{trace }\mathbf{u}\mathbf{v}^{\mathrm{T}} = \mathbf{uv} \tag{12.18}$$

$$\text{trace }\not{\mathbf{u}}\not{\mathbf{v}} = -2\mathbf{uv} \tag{12.19}$$

$$\text{trace }\not{\mathbf{u}}\mathbf{H} = \text{trace }\not{\mathbf{u}}\mathbf{H}_- = -2\mathbf{u}\not{\mathbf{H}} \tag{12.20}$$

* * *

$$|\,\mathbf{n}\,| \overset{\text{prec.}}{=} 1: \quad \mathbf{P} \overset{\text{def.}}{=} \mathbf{n}\mathbf{n}^{\mathrm{T}} \tag{12.21}$$

$$\mathbf{Q} \overset{\text{def.}}{=} \mathbf{1} - \mathbf{P} \tag{12.22}$$

$$\mathbf{P} = \mathbf{P}^{\mathrm{T}} = \mathbf{P}^2 = 1 + \not{\mathbf{n}}^2 \tag{12.23}$$

$$\mathbf{Q} = \mathbf{Q}^{\mathrm{T}} = \mathbf{Q}^2 = -\not{\mathbf{n}}^2 \tag{12.24}$$

$$\mathbf{PQ} = \mathbf{QP} = \mathbf{0} \tag{12.25}$$

$$\mathbf{u} = \underbrace{\mathbf{n}(\mathbf{nu})}_{\mathbf{Pu}} - \underbrace{\mathbf{n}\times(\mathbf{n}\times\mathbf{u})}_{\mathbf{Qu}} = \mathbf{n}(\mathbf{nu}) - \not{\mathbf{n}}^2\mathbf{u} \tag{12.26}$$

$$\cancel{\mathbf{u}} = \cancel{\mathbf{n}}\underbrace{(\mathbf{nu})}_{\mathrm{Q|Q}} - \underbrace{\mathbf{nu}^{\mathrm{T}}\cancel{\mathbf{n}}}_{\mathrm{P|Q}} - \underbrace{\cancel{\mathbf{n}}\mathbf{un}^{\mathrm{T}}}_{\mathrm{Q|P}} \tag{12.27}$$

$$\mathbf{h_n} \stackrel{\text{def}}{=} \mathbf{Hn} : \quad \left(\mathbf{H}^{\mathrm{T}} - \operatorname{trace}\mathbf{H} \cdot 1\right) \cdot \cancel{\mathbf{n}} = \cancel{\mathbf{n}} \cdot \mathbf{H} \cdot \cancel{\mathbf{n}}^2 + \cancel{\mathbf{h}}_{\mathbf{n}} \cdot \cancel{\mathbf{n}}^2 \tag{12.28}$$

13 Tensor Analysis

Abstract

The following calculation rules concern the formations of gradients, divergences and rotations.

The following calculation rules[1] concern the formations of gradients, divergences and rotations using the Nabla operator (13.1). Some rules are formulated in matrix notation, which has the advantage that one can use the rules of matrix calculus. Here, unlike usual, the differential operator must sometimes be placed to the right of the function on which it acts, as in the case of the gradient of a vector field (13.5). In such unconventional cases, the functions on which the differential operator acts are denoted by a vertical arrow. Let ψ denote a scalar field, **u** and **v** vector fields, **r** the position vector field (12.9) and **H** a tensor field.

$$\nabla \overset{\text{def}}{=} \begin{bmatrix} \partial_1 \\ \partial_2 \\ \partial_3 \end{bmatrix} = \boldsymbol{e}_i \cdot \partial_i; \quad \nabla_i = \partial_i \tag{13.1}$$

[1] Many of these rules are stated by Gurtin [2, p. 12].

P. Halfar, *Stresses in glaciers*, https://doi.org/10.1007/978-3-662-66024-9_13

$$\not\nabla = \begin{bmatrix} 0 & -\partial_3 & \partial_2 \\ \partial_3 & 0 & -\partial_1 \\ -\partial_2 & \partial_1 & 0 \end{bmatrix} = \not{e}_{\mathrm{i}} \cdot \partial_i; \quad \not\nabla_{ij} = -\epsilon_{ijk} \cdot \partial_k \tag{13.2}$$

$$\partial_n \stackrel{\text{def.}}{=} \mathbf{n} \cdot \nabla \tag{13.3}$$

* * *

$$\operatorname{grad} \psi \stackrel{\text{def.}}{=} \nabla \psi; \quad [\operatorname{grad} \psi]_i = \partial_i \psi \tag{13.4}$$

$$\operatorname{grad} \mathbf{u} \stackrel{\text{def.}}{=} \downarrow \mathbf{u} \nabla^{\mathrm{T}}; \quad [\operatorname{grad} \mathbf{u}]_{ik} = \partial_k u_i \tag{13.5}$$

$$\operatorname{rot} \mathbf{u} \stackrel{\text{def}}{=} \nabla \times \boldsymbol{u} = \not\nabla \mathbf{u}; \quad [\operatorname{rot} \mathbf{u}]_i = \epsilon_{ijk} \partial_j u_k \tag{13.6}$$

$$\operatorname{div} \mathbf{u} \stackrel{\text{def.}}{=} \nabla \mathbf{u} = \partial_i u_i \tag{13.7}$$

$$\operatorname{div} \mathbf{H} \stackrel{\text{def.}}{=} \downarrow \mathbf{H} \nabla; \quad [\operatorname{div} \mathbf{H}]_i = \partial_k H_{ik} \tag{13.8}$$

$$\operatorname{rot} \mathbf{H} \stackrel{\text{def}}{=} \not\nabla \mathbf{H}^{\mathrm{T}}; \quad [\operatorname{rot} \mathbf{H}]_{ik} = \epsilon_{ilm} \partial_l H_{km} \tag{13.9}$$

* * *

$$\operatorname{grad} \mathbf{r} = \mathbf{1} \tag{13.10}$$

$$\operatorname{trace} \operatorname{grad} \mathbf{u} = \operatorname{div} \mathbf{u} \tag{13.11}$$

$$\operatorname{grad} (\not{\mathbf{u}} \mathbf{v}) = \operatorname{grad} (\mathbf{u} \times \mathbf{v}) = \not{\mathbf{u}} \cdot \operatorname{grad} \mathbf{v} - \not{\mathbf{v}} \cdot \operatorname{grad} \mathbf{u} \tag{13.12}$$

$$\operatorname{rot} \not{\mathbf{u}} = -\operatorname{grad} \mathbf{u} + (\operatorname{div} \mathbf{u}) \cdot 1 \tag{13.13}$$

$$\operatorname{trace} \operatorname{rot} \not{\mathbf{u}} = 2 \cdot \operatorname{div} \mathbf{u} \tag{13.14}$$

$$\operatorname{rot} \not{\mathbf{u}} - \frac{1}{2} \cdot \operatorname{trace} (\operatorname{rot} \not{\mathbf{u}}) \cdot 1 = -\operatorname{grad} \mathbf{u} \tag{13.15}$$

$$\operatorname{rot} \mathbf{H} = \operatorname{rot} \mathbf{H}_{+} - \operatorname{grad} \not{\mathbf{H}} + (\operatorname{div} \not{\mathbf{H}}) \cdot 1 \tag{13.16}$$

$$(\operatorname{rot} \operatorname{rot} \mathbf{H})^T = \operatorname{rot} \operatorname{rot} \mathbf{H}^{\mathrm{T}} \tag{13.17}$$

$$(\text{rot rot } \mathbf{H})_+ = \text{rot rot } \mathbf{H}_+ \tag{13.18}$$

$$\text{trace } (\text{rot } \mathbf{H}) = 2 \cdot \text{div } \not{\mathbf{H}} \tag{13.19}$$

$$\text{rot } \mathbf{H} - \frac{1}{2} \cdot \text{trace } (\text{rot } \mathbf{H}) \cdot 1 = \text{rot } \mathbf{H}_+ - \text{grad } \not{\mathbf{H}} \tag{13.20}$$

$$\text{div } (\not{\mathbf{r}}\mathbf{H}) = \mathbf{r} \times \text{div } \mathbf{H} + 2 \cdot \not{\mathbf{H}} \tag{13.21}$$

$$\text{rot } (\not{\mathbf{r}}\mathbf{H}) = -(\text{rot } \mathbf{H}) \cdot \not{\mathbf{r}} - \mathbf{H} + \text{trace } \mathbf{H} \cdot 1. \tag{13.22}$$

* * *

$$\text{rot } \mathbf{H} = \begin{bmatrix} \partial_2 H_{13} - \partial_3 H_{12} & \partial_2 H_{23} - \partial_3 H_{22} & \partial_2 H_{33} - \partial_3 H_{32} \\ \partial_3 H_{11} - \partial_1 H_{13} & \partial_3 H_{21} - \partial_1 H_{23} & \partial_3 H_{31} - \partial_1 H_{33} \\ \partial_1 H_{12} - \partial_2 H_{11} & \partial_1 H_{22} - \partial_2 H_{21} & \partial_1 H_{32} - \partial_2 H_{31} \end{bmatrix} \tag{13.23}$$

$$\text{rot rot } \mathbf{H} = \begin{bmatrix} \begin{matrix} \partial_2^2 H_{33} + \partial_3^2 H_{22} \\ -\partial_2 \partial_3 (H_{23} + H_{32}) \end{matrix} & \begin{matrix} -\partial_1 \partial_2 H_{33} - \partial_3^2 H_{21} \\ +\partial_3(\partial_1 H_{23} + \partial_2 H_{31}) \end{matrix} & \begin{matrix} -\partial_1 \partial_3 H_{22} - \partial_2^2 H_{31} \\ +\partial_2(\partial_1 H_{32} + \partial_3 H_{21}) \end{matrix} \\ \begin{matrix} -\partial_1 \partial_2 H_{33} - \partial_3^2 H_{12} \\ +\partial_3(\partial_1 H_{32} + \partial_2 H_{13}) \end{matrix} & \begin{matrix} \partial_1^2 H_{33} + \partial_3^2 H_{11} \\ -\partial_1 \partial_3 (H_{13} + H_{31}) \end{matrix} & \begin{matrix} -\partial_2 \partial_3 H_{11} - \partial_1^2 H_{32} \\ +\partial_1(\partial_2 H_{31} + \partial_3 H_{12}) \end{matrix} \\ \begin{matrix} -\partial_1 \partial_3 H_{22} - \partial_2^2 H_{13} \\ +\partial_2(\partial_1 H_{23} + \partial_3 H_{12}) \end{matrix} & \begin{matrix} -\partial_2 \partial_3 H_{11} - \partial_1^2 H_{23} \\ +\partial_1(\partial_2 H_{13} + \partial_3 H_{21}) \end{matrix} & \begin{matrix} \partial_1^2 H_{22} + \partial_2^2 H_{11} \\ -\partial_1 \partial_2 (H_{12} + H_{21}) \end{matrix} \end{bmatrix} \tag{13.24}$$

Redundancy Functions and Normalizations 14

Abstract

Two stress functions lead to the same "weightless stress tensor field" exactly when their difference is a so-called redundancy function. The redundancy functions can be written as sums of any antisymmetric tensor field and of the gradient of any vector field.

14.1 Redundancy Functions

Two stress functions lead to the same stress tensor field if and only if their difference is a so-called redundancy function $\mathbf{A}^{\bullet}$ whose **T**-field $\mathbf{T}^{\bullet}$ vanishes (14.1). The redundancy functions $\mathbf{A}^{\bullet}$ (14.5) can be written as sums of any antisymmetric tensor field $\boldsymbol{u}$ and of the gradient of any vector field **v**.[1] The redundancy functions can also be characterized in that their symmetric part $\mathbf{A}^{\bullet}_{+}$ (14.6) is the symmetrized gradient of any vector field **v** and that their antisymmetric part $\mathbf{A}^{\bullet}_{-}$ (14.7) is arbitrary. The **B**- and **C**-fields $\mathbf{B}^{\bullet}$ (14.8) and $\mathbf{C}^{\bullet}$ (14.9) of the redundancy functions $\mathbf{A}^{\bullet}$ are gradient fields.

—

$$\mathbf{T}^{\bullet} \overset{\text{def}}{=} \operatorname{rot}\left[\operatorname{rot}\mathbf{A}^{\bullet} - \frac{1}{2}\cdot\operatorname{trace}(\operatorname{rot}\mathbf{A}^{\bullet})\cdot\mathbf{1}\right] \overset{(13.20)}{=} \operatorname{rot}\operatorname{rot}\mathbf{A}^{\bullet}_{+} \overset{\text{prec.}}{=} \mathbf{0} \qquad (14.1)$$

[1] Every expression (14.5) is a redundancy function and every redundancy function can be written like this.

P. Halfar, *Stresses in glaciers*, https://doi.org/10.1007/978-3-662-66024-9_14

* * *

$$\operatorname{rot}\mathbf{A}^{\bullet} - \frac{1}{2}\cdot \operatorname{trace}(\operatorname{rot}\mathbf{A}^{\bullet})\cdot 1 = -\operatorname{grad}\mathbf{u} \tag{14.2}$$

$$-\frac{1}{2}\cdot \operatorname{trace}(\operatorname{rot}\mathbf{A}^{\bullet}) = -\operatorname{div}\mathbf{u} \tag{14.3}$$

$$\operatorname{rot}\mathbf{A}^{\bullet} = -\operatorname{grad}\mathbf{u} + \operatorname{div}\mathbf{u}\cdot 1 \overset{(13.13)}{=} \operatorname{rot}\not\mathbf{u} \tag{14.4}$$

* * *

$$\mathbf{A}^{\bullet} = \not\mathbf{u} + \operatorname{grad}\mathbf{v} \tag{14.5}$$

$$\mathbf{A}^{\bullet}_{+} = \frac{1}{2}\left[\operatorname{grad}\mathbf{v} + (\operatorname{grad}\mathbf{v})^{\mathrm{T}}\right] \tag{14.6}$$

$$\mathbf{A}^{\bullet}_{-} = \frac{1}{2}\left[\operatorname{grad}\mathbf{v} - (\operatorname{grad}\mathbf{v})^{\mathrm{T}}\right] + \not\mathbf{u} \tag{14.7}$$

$$\mathbf{B}^{\bullet} \overset{\text{def}}{=} \operatorname{rot}\mathbf{A}^{\bullet} - \frac{1}{2}\cdot \operatorname{trace}(\operatorname{rot}\mathbf{A}^{\bullet})\cdot \mathbf{1} = -\operatorname{grad}\mathbf{u} \tag{14.8}$$

$$\mathbf{C}^{\bullet} \overset{\text{def}}{=} \mathbf{A}^{\bullet} + \not\mathbf{r}\mathbf{B}^{\bullet} = \operatorname{grad}(\mathbf{v} - \mathbf{r}\times\mathbf{u}). \tag{14.9}$$

14.2 Normalizations

In this section it is proved that all weightless stress tensor fields **T** can be obtained from tensor fields **A** (stress functions) which have one of the five following normalizations.[2] For each of these five normalizations it is shown that a symmetric redundancy function (14.6) can be added to each **A** symmetric tensor field, so that the result is a tensor field with this normalization, i.e. that the corresponding three (of the six independent) tensor components disappear.

[2] See Sect. 6.2.2.

14.2.1 xx-yy-zz Normalization

For the xx-yy-zz or 11-22-33 normalization, one must make the non-diagonal elements of the symmetric stress function matrix **A** vanish by adding a symmetric redundancy function (14.10)–(14.12).

Using the integral f of $-2A_{23}$ (14.13), these conditions can be transformed into a condition for a function ψ (14.18). Such an actually existing function ψ shows[3] that the xx-yy-zz normalization is admissible, that is, that one can obtain all weightless stress tensor fields from the stress functions normalized in this way.

$$\partial_1 v_2 + \partial_2 v_1 = -2 \cdot A_{12} \tag{14.10}$$

$$\partial_2 v_3 + \partial_3 v_2 = -2 \cdot A_{23} \tag{14.11}$$

$$\partial_3 v_1 + \partial_1 v_3 = -2 \cdot A_{31} \tag{14.12}$$

* * *

$$\partial_2 f \stackrel{\text{def.}}{=} -2 \cdot A_{23} \tag{14.13}$$

* * *

$$v_2 = \partial_2 \psi \tag{14.14}$$

$$v_3 = f - \partial_3 \psi \tag{14.15}$$

$$\partial_2 v_1 = -\partial_1 \partial_2 \psi - 2 \cdot A_{12} \tag{14.16}$$

$$\partial_3 v_1 = \partial_1 \partial_3 \psi - \partial_1 f - 2 \cdot A_{31} \tag{14.17}$$

* * *

$$\partial_1 \partial_2 \partial_3 \psi = -\partial_3 A_{12} + \partial_2 \cdot A_{31} - \partial_1 A_{23} \tag{14.18}$$

[3] If you have to go into areas outside of the definition area Ω for integrations, where at first **A** is not defined, you continue **A** into the outer area of Ω somehow.

14.2.2 The Normalizations *xx-yy-xy, xx-yy-xz, xx-xy-yz, xy-yz-xz*

To convert a stress function **A** into its respective normalized form, one determines the components of **v** in the redundancy function (14.6) successively in such a way that the corresponding components of the stress function vanish by adding the redundancy function. This is represented by the following scheme:

Normalization	v_3	v_2	v_1
$11-22-12$ or $xx-yy-xy$	$\partial_3 v_3 : A_{33} \to 3$	$\partial_3 v_2 : A_{23} \to 3$	$\partial_3 v_1 : A_{13} \to 0$
$11-22-13$ or $xx-yy-xz$	$\partial_3 v_3 : A_{33} \to 3$	$\partial_3 v_2 : A_{23} \to 3$	$\partial_2 v_1 : A_{12} \to 2$
$11-12-23$ or $xx-xy-yz$	$\partial_3 v_3 : A_{33} \to 3$	$\partial_2 v_2 : A_{22} \to 2$	$\partial_3 v_1 : A_{13} \to 3$
$12-23-13$ or $xy-yz-xz$	$\partial_3 v_3 : A_{33} \to 3$	$\partial_2 v_2 : A_{22} \to 2$	$\partial_1 v_1 : A_{11} \to 1$

For example, in the normalization *xx-yy-xy,* first $\partial_3 v_3$ and then v_3 are determined by integration in the x_3 direction such that the 33 component of the stress function becomes zero, then $\partial_3 v_2$ and v_2 are determined such that the 23 component becomes zero, and finally $\partial_3 v_1$ and v_1 are determined such that the 13 component becomes zero.

14.3 Normalizations with Boundary Conditions

Among the **A**-fields (7.14) of the general solution **T** (7.15) sets of matrix fields $\mathbf{A}_0$ play a role, which together with their first derivatives vanish on the boundary surface Σ and are otherwise arbitrary. In the following the conditions are given according to which these $\mathbf{A}_0$–fields can be normalized by additions of redundancy functions in such a way that $\mathbf{A}_0$-fields result again , i.e. that also the normalized matrix fields together with their first derivatives vanish on the boundary surface Σ of given boundary stresses.[4]

In any case, all $\mathbf{A}_0$– matrix fields can always be normalized to symmetric $\mathbf{A}_0$ – matrix fields by omitting their antisymmetric part. The normalization of all symmetric $\mathbf{A}_0$ – matrix fields to one of the five above mentioned normalization types xx-yy-zz etc. is done according to Sect. 14.2 by adding suitable symmetric redundancy functions (14.6) with the i-k components $(\partial_i v_k + \partial_k v_i)/2$. This normalization can then be made in such a way that

[4] See Sect. 7.4.

the normalized matrix fields are again $\mathbf{A}_0$– matrix fields, if one can choose the three fields v_i in such a way that they vanish together with their first and second derivatives on the boundary surface Σ, because then the redundancy functions (14.6) also vanish together with their first derivatives on the boundary surface Σ and thus also the normalized fields. So we only have to show that the three fields v_i together with their first and second derivatives vanish on the boundary surface Σ.

In the normalization procedures in Sect. 14.2 these fields v_i are determined by integration in the direction of one of the coordinate axes. If the boundary surface Σ lies[5] transverse to all directions of integration and one begins with the integrations on the boundary surface, then these v_i fields disappear on the boundary surface Σ and likewise their first and second derivatives in the direction of integration, since the $\mathbf{A}_0$– matrix fields to be normalized together with their first derivatives also disappear there. But if a function and its first and second derivatives taken in a certain direction vanish on Σ, where the direction of derivative is transverse to Σ, then this function vanishes with all its first and second derivatives on Σ. Thus it is shown that the three fields v_i vanish together with their first and second derivatives on the boundary surface Σ.

With the various normalizations, integrations occur in the following directions:

Normalization	Integration or normalization directions
xx-yy-zz	*x,y,z*
xx-yy-xy	*z*
xx-yy-xz	*y,z*
xx-xy-yz	*y,z*
xy-yz-xz	*x,y,z*

Thus, all $\mathbf{A}_0$– matrix fields can be normalized in such a way that the normalized matrix fields are $\mathbf{A}_0$– matrix fields again, if the integration directions of the respective normalization, which are also called normalization directions, are transverse to the boundary surface Σ.

[5] This means that all straight lines in the direction of integration through the boundary surface do Σ not intersect it a second time.

15 Analysis on Curved Surfaces

Abstract

In this chapter, curvilinear surface coordinates are introduced on the curved boundary surface of given boundary stresses in order to represent differential operators and boundary fields defined on this boundary surface.

15.1 Curvilinear Coordinates

In this section, curvilinear surface coordinates are introduced on the curved boundary surface Σ of given boundary stresses.

The surface Σ is defined using curvilinear surface coordinates x', y' and specifying the surface vector $\mathbf{r}_\Sigma(x', y')$, leading from the origin of the Cartesian x-y-z coordinate system to the surface Σ, as a function of these curvilinear coordinates (15.3). If these coordinates x' and y' vary, the tip of this surface vector moves on the surface Σ. The numbered denotation x'_1, x'_2 of curvilinear coordinates is used synonymously (15.1). The corresponding differential operators (15.2) denote derivatives in the direction of the curvilinear coordinate lines on the surface. By these derivatives, at each point on the surface Σ, the tangent vector basis is defined, consisting of two tangent vectors $\mathbf{f}_1$ and $\mathbf{f}_2$ (15.4), which are parallel to the curvilinear x'- and y'-coordinate lines on the surface and which span the tangent plane of the surface Σ. With the help of these tangential vectors and $\mathbf{f}_1$ and $\mathbf{f}_2$ the oriented normal vector $\mathbf{n}$ (15.5), the vectorial tangential path element $\mathbf{dr}$ (15.6) and the area element dA (15.7) can be specified on the surface Σ.

The vectors $\mathbf{f}^1$ and $\mathbf{f}^2$ of the dual tangential vector basis also span the tangential plane. With the help of this dual tangential vector basis the expansion according to the tangential vector basis and according to the normal vector can be given for a vector field $\mathbf{u}$ on the

P. Halfar, *Stresses in glaciers*, https://doi.org/10.1007/978-3-662-66024-9_15

surface Σ or alternatively the expansion according to the dual tangential vector basis and according to the normal vector (15.8)–(15.11).

Using these tangent vector bases and the normal vector, at any point on the surface Σ, the projector **P** onto the normal vector, the projector **Q** onto the tangent plane, and the skew-symmetric tensor **n** can be computed (15.12)–(15.14). Choosing the Cartesian coordinate system so that the third axis points in the direction of the normal vector, we obtain simple matrix representations for these quantities (15.15).

If the surface Σ can be represented by a function $z = z_0(x, y)$, the curvilinear coordinate grid on that surface can be generated from the Cartesian coordinates x and y by projecting these coordinates parallel to the z axis onto the surface (15.16)–(15.21).

—

$$x'_1 = x'; \quad x'_2 = y' \tag{15.1}$$

$$\partial'_x = \partial'_1 = \partial/\partial x'; \quad \partial'_y = \partial'_2 = \partial/\partial y' \tag{15.2}$$

$$\mathbf{r}_\Sigma(x', y') = \begin{bmatrix} x_\Sigma(x', y') \\ y_\Sigma(x', y') \\ z_\Sigma(x', y') \end{bmatrix} = \begin{bmatrix} x_{1\Sigma}(x'_1, x'_2) \\ x_{2\Sigma}(x'_1, x'_2) \\ x_{3\Sigma}(x'_1, x'_2) \end{bmatrix} \tag{15.3}$$

$$\mathbf{f}_1 = \partial'_x \mathbf{r}_\Sigma(x', y') = \begin{bmatrix} \partial'_x x_\Sigma(x', y') \\ \partial'_x y_\Sigma(x', y') \\ \partial'_x z_\Sigma(x', y') \end{bmatrix}; \quad \mathbf{f}_2 = \partial'_y \mathbf{r}_\Sigma(x', y') = \begin{bmatrix} \partial'_y x_\Sigma(x', y') \\ \partial'_y y_\Sigma(x', y') \\ \partial'_y z_\Sigma(x', y') \end{bmatrix} \tag{15.4}$$

$$\mathbf{n} = \frac{\mathbf{f}_1 \times \mathbf{f}_2}{|\mathbf{f}_1 \times \mathbf{f}_2|} \tag{15.5}$$

$$\mathbf{dr} = \partial'_x \mathbf{r}_\Sigma \cdot dx' + \partial'_y \mathbf{r}_\Sigma \cdot dy' = \mathbf{f}_1 \cdot dx' + \mathbf{f}_2 \cdot dy' \tag{15.6}$$

$$dA = |\mathbf{f}_1 \cdot dx' \times \mathbf{f}_2 \cdot dy'| = |\mathbf{f}_1 \times \mathbf{f}_2| \cdot dx' \cdot dy' \tag{15.7}$$

* * *

$$\mathbf{f}^1 = -\frac{\mathbf{n} \times \mathbf{f}_2}{|\mathbf{f}_1 \times \mathbf{f}_2|}; \quad \mathbf{f}^2 = \frac{\mathbf{n} \times \mathbf{f}_1}{|\mathbf{f}_1 \times \mathbf{f}_2|} \tag{15.8}$$

$$\mathbf{f}^\mu \mathbf{f}_\nu = \delta_{\mu\nu}; \quad \mu, \nu = 1, 2 \tag{15.9}$$

$$\mathbf{n} \times \mathbf{f}^1 = \frac{\mathbf{f}_2}{|\mathbf{f}_1 \times \mathbf{f}_2|}; \quad \mathbf{n} \times \mathbf{f}^2 = -\frac{\mathbf{f}_1}{|\mathbf{f}_1 \times \mathbf{f}_2|} \tag{15.10}$$

$$\mathbf{u} = \left(\mathbf{u}\mathbf{f}^1\right) \cdot \mathbf{f}_1 + \left(\mathbf{u}\mathbf{f}^2\right) \cdot \mathbf{f}_2 + (\mathbf{u}\mathbf{n}) \cdot \mathbf{n} = (\mathbf{u}\mathbf{f}_1) \cdot \mathbf{f}^1 + (\mathbf{u}\mathbf{f}_2) \cdot \mathbf{f}^2 + (\mathbf{u}\mathbf{n}) \cdot \mathbf{n} \tag{15.11}$$

* * *

$$\mathbf{P} = \mathbf{n}\mathbf{n}^{\mathrm{T}} \tag{15.12}$$

$$\mathbf{Q} = \mathbf{f}^{\mu}\mathbf{f}_{\mu}^{\mathrm{T}} = \mathbf{f}_{\mu}\mathbf{f}^{\mu\mathrm{T}} \tag{15.13}$$

$$\begin{aligned} \not{\mathbf{n}} &= \left(\mathbf{n} \times \mathbf{f}_{\mu}\right) \cdot \mathbf{f}^{\mu\mathrm{T}} = |\mathbf{f}_1 \times \mathbf{f}_2| \cdot \left(-\mathbf{f}^1\mathbf{f}^{2\mathrm{T}} + \mathbf{f}^2\mathbf{f}^{1\mathrm{T}}\right) \\ &= \left(\mathbf{n} \times \mathbf{f}^{\mu}\right) \cdot \mathbf{f}_{\mu}^{\mathrm{T}} = |\mathbf{f}_1 \times \mathbf{f}_2|^{-1} \cdot \left(-\mathbf{f}_1\mathbf{f}_2^{\mathrm{T}} + \mathbf{f}_2\mathbf{f}_1^{\mathrm{T}}\right) \end{aligned} \tag{15.14}$$

$$\mathbf{n} = \begin{bmatrix} 0 \\ 0 \\ 1 \end{bmatrix} : \quad \mathbf{P} = \begin{bmatrix} 0 & 0 & 0 \\ 0 & 0 & 0 \\ 0 & 0 & 1 \end{bmatrix}; \mathbf{Q} = \begin{bmatrix} 1 & 0 & 0 \\ 0 & 1 & 0 \\ 0 & 0 & 0 \end{bmatrix}; \not{\mathbf{n}} = \begin{bmatrix} 0 & -1 & 0 \\ 1 & 0 & 0 \\ 0 & 0 & 0 \end{bmatrix} \tag{15.15}$$

* * *

$$\mathbf{r}_{\Sigma}(x', y') = \mathbf{r}_0(x', y') = \begin{bmatrix} x' \\ y' \\ z_0(x', y') \end{bmatrix} \tag{15.16}$$

$$\mathbf{f}_1 = \partial'_x \mathbf{r}_0 = \begin{bmatrix} 1 \\ 0 \\ \partial'_x z_0 \end{bmatrix}; \quad \mathbf{f}_2 = \partial'_y \mathbf{r}_0 = \begin{bmatrix} 0 \\ 1 \\ \partial'_y z_0 \end{bmatrix} \tag{15.17}$$

$$\mathbf{f}_1 \times \mathbf{f}_2 = \begin{bmatrix} -\partial'_x z_0 \\ -\partial'_y z_0 \\ 1 \end{bmatrix}; \quad |\,\mathbf{f}_1 \times \mathbf{f}_2\,| = \left[1 + \left(\partial'_x z_0\right)^2 + \left(\partial'_y z_0\right)^2\right]^{1/2} \tag{15.18}$$

$$\mathbf{n} = \left[1 + \left(\partial'_x z_0\right)^2 + \left(\partial'_y z_0\right)^2\right]^{-1/2} \begin{bmatrix} -\partial'_x z_0 \\ -\partial'_y z_0 \\ 1 \end{bmatrix} \tag{15.19}$$

$$\mathbf{f}^1 = \left[1 + \left(\partial'_x z_0\right)^2 + \left(\partial'_y z_0\right)^2\right]^{-1} \begin{bmatrix} 1 + \left(\partial'_y z_0\right)^2 \\ -\partial'_x z_0 \cdot \partial'_y z_0 \\ \partial'_x z_0 \end{bmatrix} \tag{15.20}$$

$$\mathbf{f}^2 = \left[1 + \left(\partial'_x z_0\right)^2 + \left(\partial'_y z_0\right)^2\right]^{-1} \begin{bmatrix} -\partial'_x z_0 \cdot \partial'_y z_0 \\ 1 + \left(\partial'_x z_0\right)^2 \\ \partial'_y z_0 \end{bmatrix} \tag{15.21}$$

15.2 Differential Operators and Derivatives

In the boundary conditions, linear differential operators occur on the boundary surface Σ. In this section, the calculation rules for these differential operators are presented.

These differential operators are generally tensor or matrix operators. The scalar operator ∂_n causes a derivative in the direction of the oriented normal $\mathbf{n}$ on the boundary surface Σ. The gradient operator ∇ as well as the operator $\mathbf{n} \times \nabla$ are single column matrix operators and the rotation operator $\nabla\!\!\!\!\times$ is a quadratic matrix operator. These matrix operators can be expressed by the normal derivative ∂_n and by the derivatives ∂_1' and ∂_2' according to the curvilinear surface coordinates. The projections of these operators onto the normal direction and the tangent plane are obtained by multiplying by the corresponding projectors $\mathbf{P}$ or $\mathbf{Q}$. All functions appearing in these operator expressions (15.22)–(15.29) are not differentiated, so they are to be treated as constants.[1]

The matrix operators are applied by matrix multiplication to scalar functions ψ, to single-column matrix functions (vector functions) $\mathbf{u}$ or to quadratic matrix functions (tensor functions) $\mathbf{A}$ (15.30)–(15.39). In doing so, the operators can, unlike usual, also act to the left. In such deviations from the usual conventions, the functions on which the operators act are marked by a vertical arrow.

Some operators contain derivatives acting only tangentially to the surface Σ and can therefore be applied to functions defined only on the surface Σ. Thus the projection of the gradient vector onto the tangent plane (15.25) contains only tangentially acting derivatives, so that for a scalar function ψ the gradient field of this function is generated on the surface Σ (15.32). This gradient field is tangential to the surface and is perpendicular to the level lines of constant ψ-values on the surface Σ. The vector operator $\mathbf{n} \times \nabla$ (15.26) also contains only tangential derivatives and generates from the scalar function ψ the tangential gradient field rotated by a right angle, which is consequently parallel to the level lines of constant ψ-values (15.33). The rotation operator projected onto the normal direction $\nabla\!\!\!\!\times \cdot \mathbf{P}$ also contains only tangential derivatives (15.28) and (15.38).

—

$$\partial_n \overset{\text{def.}}{=} \mathbf{n} \cdot \nabla \tag{15.22}$$

$$\nabla = \mathbf{f}^1 \cdot \partial_1' + \mathbf{f}^2 \cdot \partial_2' + \mathbf{n} \cdot \partial_n \tag{15.23}$$

$$\mathbf{P} \cdot \nabla = \mathbf{n} \cdot \partial_n \tag{15.24}$$

[1] These constants are given by the corresponding function values at the point under consideration.

$$\mathbf{Q}\cdot\nabla=\mathbf{f}^1\cdot\partial_1'+\mathbf{f}^2\cdot\partial_2' \tag{15.25}$$

$$\mathbf{n}\times\nabla=\not{\mathbf{n}}\cdot\nabla=\frac{1}{|\mathbf{f}_1\times\mathbf{f}_2|}\cdot\left(\mathbf{f}_2\cdot\partial_1'-\mathbf{f}_1\cdot\partial_2'\right). \tag{15.26}$$

$$\not{\nabla}\overset{(12.27)}{=}\underbrace{\not{\mathbf{n}}\partial_n}_{Q\not{\nabla}Q}-\underbrace{\mathbf{n}\nabla^{\mathrm{T}}\not{\mathbf{n}}}_{P\not{\nabla}Q}-\underbrace{\not{\mathbf{n}}\nabla\mathbf{n}^{\mathrm{T}}}_{Q\not{\nabla}P}=\not{\mathbf{n}}\cdot\partial_n+\frac{1}{|\mathbf{f}_1\times\mathbf{f}_2|}\left[\left(\mathbf{n}\mathbf{f}_2^{\mathrm{T}}-\mathbf{f}_2\mathbf{n}^{\mathrm{T}}\right)\cdot\partial_1'-\left(\mathbf{n}\mathbf{f}_1^{\mathrm{T}}-\mathbf{f}_1\mathbf{n}^{\mathrm{T}}\right)\cdot\partial_2'\right] \tag{15.27}$$

$$\not{\nabla}\mathbf{P}=-\not{\mathbf{n}}\nabla\mathbf{n}^{\mathrm{T}}=+\frac{1}{|\mathbf{f}_1\times\mathbf{f}_2|}\left[-\mathbf{f}_2\mathbf{n}^{\mathrm{T}}\cdot\partial_1'+\mathbf{f}_1\mathbf{n}^{\mathrm{T}}\cdot\partial_2'\right] \tag{15.28}$$

$$\not{\nabla}\mathbf{Q}=\not{\mathbf{n}}\partial_{\mathrm{n}}-\mathbf{n}\nabla^{\mathrm{T}}\not{\mathbf{n}}=\not{\mathbf{n}}\cdot\partial_n+\frac{1}{|\mathbf{f}_1\times\mathbf{f}_2|}\left[\mathbf{n}\mathbf{f}_2^{\mathrm{T}}\cdot\partial_1'-\mathbf{n}\mathbf{f}_1^{\mathrm{T}}\cdot\partial_2'\right] \tag{15.29}$$

* * *

$$\nabla\psi=\mathbf{f}^1\cdot\partial_1'\psi+\mathbf{f}^2\cdot\partial_2'\psi+\mathbf{n}\cdot\partial_n\psi \tag{15.30}$$

$$\mathbf{P}\cdot\nabla\psi=\mathbf{n}\cdot\partial_n\psi \tag{15.31}$$

$$\mathbf{Q}\cdot\nabla\psi=\mathbf{f}^1\cdot\partial_1'\psi+\mathbf{f}^2\cdot\partial_2'\psi \tag{15.32}$$

$$\mathbf{n}\times\nabla\psi=\frac{1}{|\ \mathbf{f}_1\times\mathbf{f}_2\ |}\cdot\left(\mathbf{f}_2\cdot\partial_1'\psi-\mathbf{f}_1\cdot\partial_2'\psi\right) \tag{15.33}$$

$$\operatorname{grad}\mathbf{u}=\downarrow\mathbf{u}\nabla^{\mathrm{T}}=\partial_1'\mathbf{u}\cdot\mathbf{f}^{1\mathrm{T}}+\partial_2'\mathbf{u}\cdot\mathbf{f}^{2\mathrm{T}}+\partial_n\mathbf{u}\cdot\mathbf{n}^{\mathrm{T}} \tag{15.34}$$

$$\operatorname{grad}\mathbf{u}\cdot\mathbf{P}=\downarrow\mathbf{u}\nabla^{\mathrm{T}}\mathbf{P}=\partial_n\mathbf{u}\cdot\mathbf{n}^{\mathrm{T}} \tag{15.35}$$

$$\operatorname{grad}\mathbf{u}\cdot\mathbf{Q}=\downarrow\mathbf{u}\nabla^{\mathrm{T}}\mathbf{Q}=\partial_1'\mathbf{u}\cdot\mathbf{f}^{1\mathrm{T}}+\partial_2'\mathbf{u}\cdot\mathbf{f}^{2\mathrm{T}} \tag{15.36}$$

$$\begin{aligned}(\operatorname{rot}\mathbf{A})^{\mathrm{T}}&=-\overset{\downarrow}{\mathbf{A}}\cdot\not{\nabla}=-\partial_n\mathbf{A}\cdot\not{\mathbf{n}}+\overset{\downarrow}{\mathbf{A}}\,\mathbf{n}\nabla^{\mathrm{T}}\not{\mathbf{n}}+\overset{\downarrow}{\mathbf{A}}\,\not{\mathbf{n}}\nabla\mathbf{n}^{\mathrm{T}}\\&=-\partial_n\mathbf{A}\cdot\not{\mathbf{n}}-\frac{1}{|\mathbf{f}_1\times\mathbf{f}_2|}\left[\partial_1'\mathbf{A}\cdot\left(\mathbf{n}\mathbf{f}_2^{\mathrm{T}}-\mathbf{f}_2\mathbf{n}^{\mathrm{T}}\right)-\partial_2'\mathbf{A}\cdot\left(\mathbf{n}\mathbf{f}_1^{\mathrm{T}}-\mathbf{f}_1\mathbf{n}^{\mathrm{T}}\right)\right]\end{aligned} \tag{15.37}$$

$$\begin{aligned}(\text{rot}\,\mathbf{A})^{\mathrm{T}} \cdot \mathbf{P} &= -\overset{\downarrow}{\mathbf{A}} \cdot \not\nabla \cdot \mathbf{P} = \overset{\downarrow}{\mathbf{A}} \not{n} \nabla \mathbf{n}^{\mathrm{T}} \\ &= -\frac{1}{|\mathbf{f}_1 \times \mathbf{f}_2|}[-\partial_1^{'} \mathbf{A} \cdot \mathbf{f}_2 \mathbf{n}^{\mathrm{T}} + \partial_2^{'} \mathbf{A} \cdot \mathbf{f}_1 \mathbf{n}^{\mathrm{T}}] \\ &\overset{(15.4)}{=} -\frac{1}{|\mathbf{f}_1 \times \mathbf{f}_2|}[-\partial_1^{'} (\mathbf{A}\mathbf{f}_2) \cdot \mathbf{n}^{\mathrm{T}} + \partial_2^{'} (\mathbf{A}\mathbf{f}_1) \cdot \mathbf{n}^{\mathrm{T}}]\end{aligned} \tag{15.38}$$

$$\begin{aligned}(\text{rot}\,\mathbf{A})^{\mathrm{T}} \cdot \mathbf{Q} &= -\overset{\downarrow}{\mathbf{A}} \cdot \not\nabla \cdot \mathbf{Q} = -\partial_n \mathbf{A} \cdot \not{n} + \overset{\downarrow}{\mathbf{A}} \mathbf{n} \nabla^{\mathrm{T}} \not{n} \\ &= -\partial_n \mathbf{A} \cdot \not{n} - \frac{1}{|\mathbf{f}_1 \times \mathbf{f}_2|}\left[\partial_1^{'} \mathbf{A} \cdot \mathbf{n}\mathbf{f}_2^{\mathrm{T}} - \partial_2^{'} \mathbf{A} \cdot \mathbf{n}\mathbf{f}_1^{\mathrm{T}}\right]\end{aligned} \tag{15.39}$$

15.3 The Boundary Fields

For the construction of the general solution one needs the $\mathbf{A}$-$\partial_n\mathbf{A}$-boundary fields $\mathbf{A}_\Sigma$ and $\partial_n\mathbf{A}$ as well as the **B**- and **C**-boundary fields resp. $\mathbf{B}_\Sigma$ and $\mathbf{C}_\Sigma$, which fit to the boundary stresses **t** given on the boundary surface Σ. These boundary fields will be characterized in the following.[2]

The matching boundary fields $\mathbf{B}_\Sigma$ resp. $\mathbf{C}_\Sigma$ are characterized by the fact that their orbital integrals over the boundary curves of partial surfaces of the boundary surface Σ must coincide with the forces resp. torques on these partial surfaces, (7.10) and (7.11) which are generated by the boundary stresses **t**. According to Stokes' theorem, this can also be expressed by differential equations (15.40) and (15.41).[3]Thus also the descent condition (15.42) for the matching boundary fields $\mathbf{B}_\Sigma$ and $\mathbf{C}_\Sigma$ is satisfied, which is a criterion for whether these boundary fields descend from a **A**-field.[4]

The matching $\mathbf{A}$-$\partial_n\mathbf{A}$-boundary fields are characterized by the fact that they lead to matching $\mathbf{B}_\Sigma - \mathbf{C}_\Sigma$ boundary fields. The corresponding condition (15.43) for the $\mathbf{A}$-$\partial_n\mathbf{A}$-boundary fields[5] can be transformed[6] to obtain both condition (15.48) for the tangential

[2]How the following expressions can be given as functions of curvilinear surface coordinates is shown in Sect. 15.4.

[3]In these differential equations (15.40) and (15.41) only tangential differential operators appear, so one can apply them to the boundary fields $\mathbf{B}_\Sigma$ and $\mathbf{C}_\Sigma$.

[4]In this descent condition (6.16) only the projection part (15.42) obtained by multiplication with **n** plays a role here. Only this projection part of the descent condition contains purely tangential differential operators and these can be applied to the boundary fields $\mathbf{B}_\Sigma$ and $\mathbf{C}_\Sigma$.

[5]The matrix field $(\text{rot}\,\mathbf{A})_\Sigma^T$ occurring in condition (15.43) is defined by the $\mathbf{A}$-$\partial_n\mathbf{A}$-boundary field, so (15.43) is a condition for this boundary field.

[6]Splitting condition (15.43) into two conditions (15.44) and (15.45) is equivalent to projection decomposition by multiplying right by the projectors **P**respectively **Q**, since $\mathbf{P} = \mathbf{n}\mathbf{n}^T$ and $\mathbf{Q} = -\mathbf{n}^2$ holds.

derivatives of the boundary fields $\mathbf{A}_\Sigma$ and a condition partially fixing the normal derivatives $\partial_n\mathbf{A}$ (15.49). The part (15.50) of the normal derivatives that is not fixed can be written using any column matrix field $\mathbf{k}$ defined on the boundary surface Σ. Thus (15.48)–(15.51) there is a complete characterization of the matching $\mathbf{A}$-$\partial_n\mathbf{A}$-boundary fields.[7] For concrete calculations, two differential equations (15.40) and (15.48) must be solved for the matching boundary fields $\mathbf{B}_\Sigma$ and $\mathbf{A}_\Sigma$. The boundary fields $\mathbf{C}_\Sigma$ do not appear in the process.

One can also proceed differently, by adding the matching boundary fields $\mathbf{C}_\Sigma$ and characterizing the matching $\mathbf{A}$-$\partial_n\mathbf{A}$-boundary fields by the fact that they lead to matching boundary fields $\mathbf{B}_\Sigma$ and $\mathbf{C}_\Sigma$. In this case the boundary fields $\mathbf{A}_\Sigma$ are expressed by these boundary fields $\mathbf{B}_\Sigma$ and $\mathbf{C}_\Sigma$ (15.52). Thus the condition (15.48) for the tangential derivatives of the boundary fields $\mathbf{A}_\Sigma$ is already satisfied (15.53), because the boundary fields $\mathbf{B}_\Sigma$ and $\mathbf{C}_\Sigma$ satisfy the descent condition (15.42). Only the condition for the normal derivatives (15.49) remains. In this way, one obtains a complete characterization of the matching $\mathbf{A}$-$\partial_n\mathbf{A}$-boundary fields (15.57)–(15.60) based on matching boundary fields $\mathbf{B}_\Sigma$ and $\mathbf{C}_\Sigma$.[8] For concrete calculations, two differential equations (15.40) and (15.41) for $\mathbf{B}_\Sigma$ and $\mathbf{C}_\Sigma$ still need to be solved.

These $\mathbf{A}$-$\partial_n\mathbf{A}$-boundary fields matching a $\mathbf{B}$-$\mathbf{C}$-boundary field are listed in the same column in the structure diagram of the general solution (see Fig. 7.2) and these $\mathbf{A}$-$\partial_n\mathbf{A}$-boundary fields (15.57)–(15.60) result from arbitrary variations of the column matrix field $\mathbf{k}$. In this representation (15.57)–(15.60) of the $\mathbf{A}$-$\partial_n\mathbf{A}$-boundary fields, the dependencies on the projection components $\mathbf{B}_\Sigma\mathbf{Q}$, $\mathbf{C}_\Sigma\mathbf{Q}$, $\boldsymbol{B}_\Sigma\boldsymbol{n}$ and $\mathbf{C}_\Sigma\mathbf{n}$ of the fields $\mathbf{B}_\Sigma$ and $\mathbf{C}_\Sigma$ can be seen. Since for the construction of the general solution one needs only any suitable $\mathbf{A}$-$\partial_n\mathbf{A}$-boundary field,[9] one can set in the expressions (15.57)–(15.60) for the $\mathbf{A}$-$\partial_n\mathbf{A}$-boundary fields both the normal components $\mathbf{B}_\Sigma\mathbf{n}$ and $\mathbf{C}_\Sigma\mathbf{n}$ and also $\mathbf{k}$ zero.

The following is a proof of the claim made in Sect. 7.2 that different $\mathbf{A}$-$\partial_n\mathbf{A}$-boundary fields (15.57)–(15.60) leading to the same $\mathbf{B}$-$\mathbf{C}$-boundary field are in the same equivalence class. Thus it is to be shown that the difference of two such $\mathbf{A}$-$\partial_n\mathbf{A}$-boundary fields can be represented as the $\mathbf{A}^\bullet$-$\partial_n\mathbf{A}^\bullet$-boundary field of a redundancy function $\mathbf{A}^\bullet$, that is, that there exists a $\mathbf{A}^\bullet$-$\partial_n\mathbf{A}^\bullet$-boundary field satisfying three conditions (15.61):

The boundary field $\boldsymbol{A}^\bullet_\Sigma$ disappears.

- The projection part $\partial_n\mathbf{A}^\bullet \cdot \mathbf{Q}$ of the normal derivative disappears.
- The projection part $\partial_n\mathbf{A}^\bullet \cdot \mathbf{P}$ of the normal derivative is of the form $\mathbf{k} \cdot \mathbf{n}^\mathrm{T}$.

If we represent the redundancy function $\mathbf{A}^\bullet$ we are looking for as a gradient field (15.62) of a vector field $\mathbf{u}$, then the first of these three conditions (15.61) means that the first

[7] In equations (15.48)–(15.51) and also in the following equations (15.57)–(15.60), apart from ∂_n, only differential operators acting tangentially to the surface Σ occur.

[8] See footnote 7.

[9] This applies separately to each class (large column in Fig. 7.2) if multiple coherence occurs on the boundary surface Σ.

derivatives of this vector field must vanish on the boundary surface Σ (15.63). Then the tangential derivatives of these first derivatives also vanish and so the second condition is also satisfied (15.64). According to the third condition, (15.65) the second derivative of the vector field **u** in the normal direction must be given by the vector field **k**. Vector fields **u** with these properties (15.66) exist. Thus the assertion is proved.

$$(\mathrm{rot}\,\mathbf{B})_{\Sigma}^{\mathrm{T}} \cdot \mathbf{n} \stackrel{\mathrm{def}}{=} \overset{\downarrow}{\mathbf{B}}_{\Sigma} \cdot (\mathbf{n} \times \nabla) = \mathbf{t} \tag{15.40}$$

$$(\mathrm{rot}\,\mathbf{C})_{\Sigma}^{\mathrm{T}} \cdot \mathbf{n} \stackrel{\mathrm{def}}{=} \overset{\downarrow}{\mathbf{C}}_{\Sigma} \cdot (\mathbf{n} \times \nabla) = \mathbf{r}_{\Sigma} \times \mathbf{t} \tag{15.41}$$

* * *

$$(\mathrm{rot}\,\mathbf{C})_{\Sigma}^{\mathrm{T}} \cdot \mathbf{n} = \not{\mathbf{r}}_{\Sigma} \cdot (\mathrm{rot}\,\mathbf{B})_{\Sigma}^{\mathrm{T}} \cdot \mathbf{n} \tag{15.42}$$

* * *

$$\mathbf{B}_{\Sigma}^{\mathrm{T}} = (\mathrm{rot}\,\mathbf{A})_{\Sigma}^{\mathrm{T}} - \mathbf{1} \cdot \frac{1}{2} \cdot \mathrm{trace}\,(\mathrm{rot}\,\mathbf{A})_{\Sigma}^{\mathrm{T}} \Leftrightarrow (\mathrm{rot}\,\mathbf{A})_{\Sigma}^{\mathrm{T}} = \mathbf{B}_{\Sigma}^{\mathrm{T}} - \mathbf{1} \cdot \mathrm{trace}\,\mathbf{B}_{\Sigma} \tag{15.43}$$

$$(\mathrm{rot}\,\mathbf{A})_{\Sigma}^{\mathrm{T}} \cdot \mathbf{n} = (\mathbf{B}_{\Sigma}^{\mathrm{T}} - \mathbf{1} \cdot \mathrm{trace}\,\mathbf{B}_{\Sigma}) \cdot \mathbf{n} \tag{15.44}$$

$$(\mathrm{rot}\,\mathbf{A})_{\Sigma}^{\mathrm{T}} \cdot \mathbf{n} = \left(\mathbf{B}_{\Sigma}^{\mathrm{T}} - 1 \cdot \mathrm{trace}\,\mathbf{B}_{\Sigma}\right) \cdot \mathbf{n} \tag{15.45}$$

* * *

$$(\mathrm{rot}\,\mathbf{A})_{\Sigma}^{\mathrm{T}} \cdot \mathbf{n} \stackrel{\mathrm{id.}}{=} \overset{\downarrow}{\mathbf{A}}_{\Sigma} \cdot (\mathbf{n} \times \nabla) = \left(\mathbf{B}_{\Sigma}^{\mathrm{T}} - \mathbf{1} \cdot \mathrm{trace}\,\mathbf{B}_{\Sigma}\right) \cdot \mathbf{n} \tag{15.46}$$

$$(\mathrm{rot}\mathbf{A})_{\Sigma}^{\mathrm{T}} \cdot \not{\mathbf{n}} \stackrel{\mathrm{id.}}{=} \partial_n \mathbf{A} \cdot \mathbf{Q} - \overset{\downarrow}{\mathbf{A}}_{\Sigma} \cdot \mathbf{n} \cdot \nabla^{\mathrm{T}} \mathbf{Q} \stackrel{\mathrm{id.}}{=} \partial_n \mathbf{A} \cdot \mathbf{Q} - \left(\overset{\downarrow}{\mathbf{A}}_{\Sigma} \cdot \overset{\downarrow}{\mathbf{n}}\right) \cdot \nabla^{\mathrm{T}} \mathbf{Q} + \mathbf{A}_{\Sigma} \cdot \left(\overset{\downarrow}{\mathbf{n}} \cdot \nabla^{\mathrm{T}} \mathbf{Q}\right) = \left(\mathbf{B}_{\Sigma}^{\mathrm{T}} - 1 \cdot \mathrm{trace}\mathbf{B}_{\Sigma}\right) \cdot \not{\mathbf{n}} \tag{15.47}$$

* * *

$$\overset{\downarrow}{\mathbf{A}}_{\Sigma} \cdot (\mathbf{n} \times \nabla) = \left(\mathbf{B}_{\Sigma}^{\mathrm{T}} - \mathbf{1} \cdot \mathrm{trace}\,\mathbf{B}_{\Sigma}\right) \cdot \mathbf{n} \tag{15.48}$$

$$\partial_{\mathrm{n}} \mathbf{A} \cdot \mathbf{Q} = \left(\overset{\downarrow}{\mathbf{A}}_{\Sigma} \cdot \overset{\downarrow}{\mathbf{n}}\right) \cdot \nabla^{\mathrm{T}} \mathbf{Q} - \mathbf{A}_{\Sigma} \cdot \left(\overset{\downarrow}{\mathbf{n}} \cdot \nabla^{\mathrm{T}} \mathbf{Q}\right) + \left(\mathbf{B}_{\Sigma}^{\mathrm{T}} - 1 \cdot \mathrm{trace}\,\mathbf{B}_{\Sigma}\right) \cdot \not{\mathbf{n}} \tag{15.49}$$

$$\partial_n\mathbf{A}\cdot\mathbf{P}=\mathbf{k}\cdot\mathbf{n}^{\mathrm{T}} \tag{15.50}$$

$$\partial_n\mathbf{A}=\partial_n\mathbf{A}\cdot\mathbf{Q}+\partial_n\mathbf{A}\cdot\mathbf{P} \tag{15.51}$$

* * *

$$\mathbf{A}_\Sigma=\mathbf{C}_\Sigma-\boldsymbol{r}_\Sigma\cdot\mathbf{B}_\Sigma \tag{15.52}$$

$$\begin{aligned}\overset{\downarrow}{\mathbf{A}}_\Sigma\cdot(\mathbf{n}\times\nabla)&=\left(\overset{\downarrow}{\mathbf{C}}_\Sigma-\overset{\downarrow}{\boldsymbol{r}}_\Sigma\cdot\overset{\downarrow}{\mathbf{B}}_\Sigma\right)\cdot(\mathbf{n}\times\nabla)\\&\overset{\mathrm{id.}(13.22)}{=}\underbrace{\overset{\downarrow}{\mathbf{C}}_\Sigma\cdot(\mathbf{n}\times\nabla)-\boldsymbol{r}_\Sigma\cdot\overset{\downarrow}{\mathbf{B}}_\Sigma\cdot(\mathbf{n}\times\nabla)}_{=0;\,(15.42)}+\left[\mathbf{B}_\Sigma^{\mathrm{T}}-1\cdot\operatorname{trace}\mathbf{B}_\Sigma\right]\cdot\mathbf{n}\end{aligned} \tag{15.53}$$

$$\overset{\downarrow}{\mathbf{A}}_\Sigma\cdot\overset{\downarrow}{\mathbf{n}}\cdot\nabla^{\mathrm{T}}\mathbf{Q}\overset{(15.52)}{=}\overset{\downarrow}{\mathbf{C}}_\Sigma\cdot\overset{\downarrow}{\mathbf{n}}\cdot\nabla^{\mathrm{T}}\mathbf{Q}-\not\boldsymbol{r}_\Sigma\cdot\overset{\downarrow}{\mathbf{B}}_\Sigma\cdot\overset{\downarrow}{\mathbf{n}}\cdot\nabla^{\mathrm{T}}\mathbf{Q}-\overset{\downarrow}{\not\boldsymbol{r}}_\Sigma\cdot\mathbf{B}_\Sigma\cdot\mathbf{n}\cdot\nabla^{\mathrm{T}}\mathbf{Q} \tag{15.54}$$

$$\begin{aligned}-\overset{\downarrow}{\boldsymbol{r}}_\Sigma\cdot\overbrace{\mathbf{B}_\Sigma\cdot\mathbf{n}}^{\mathbf{b}_n}\cdot\nabla^{\mathrm{T}}\mathbf{Q}&=\not\mathbf{b}_n\cdot\overset{\downarrow}{\boldsymbol{r}}_\Sigma\cdot\nabla^{\mathrm{T}}\mathbf{Q}=\not\mathbf{b}_n\cdot\mathbf{Q}\\&\overset{(12.28)}{=}-\not\mathbf{n}\cdot\mathbf{B}_\Sigma\cdot\mathbf{Q}-\left(\mathbf{B}_\Sigma^{\mathrm{T}}-1\cdot\operatorname{trace}\mathbf{B}_\Sigma\right)\cdot\not\mathbf{n}\end{aligned} \tag{15.55}$$

$$\begin{aligned}\left(\overset{\downarrow}{\mathbf{A}}_\Sigma\cdot\overset{\downarrow}{\mathbf{n}}\right)\cdot\nabla^{\mathrm{T}}\mathbf{Q}+\left(\mathbf{B}_\Sigma^{\mathrm{T}}-1\cdot\operatorname{trace}\mathbf{B}_\Sigma\right)\cdot\not\mathbf{n}&=\overset{\downarrow}{\mathbf{C}}_\Sigma\cdot\overset{\downarrow}{\mathbf{n}}\cdot\nabla^{\mathrm{T}}\mathbf{Q}-\not\boldsymbol{r}_\Sigma\cdot\overset{\downarrow}{\mathbf{B}}_\Sigma\cdot\overset{\downarrow}{\mathbf{n}}\cdot\nabla^{\mathrm{T}}\mathbf{Q}\\&-\not\mathbf{n}\cdot\mathbf{B}_\Sigma\cdot\mathbf{Q}\end{aligned} \tag{15.56}$$

* * *

$$\mathbf{A}_\Sigma=\mathbf{C}_\Sigma-\boldsymbol{r}_\Sigma\cdot\mathbf{B}_\Sigma=\underbrace{(\mathbf{C}_\Sigma-\boldsymbol{r}_\Sigma\cdot\mathbf{B}_\Sigma)\cdot\mathbf{Q}}_{\mathbf{A}_\Sigma\cdot\mathbf{Q}}+\underbrace{\overbrace{(\mathbf{C}_\Sigma-\boldsymbol{r}_\Sigma\cdot\mathbf{B}_\Sigma)\cdot\mathbf{n}}^{\mathbf{A}_\Sigma\dot{\mathbf{n}}}\cdot\mathbf{n}^{\mathrm{T}}}_{\mathbf{A}_\Sigma\cdot\mathbf{P}} \tag{15.57}$$

$$\begin{aligned}\partial_n\mathbf{A}\cdot\mathbf{Q}=&-\underbrace{(\mathbf{C}_\Sigma-\boldsymbol{r}_\Sigma\cdot\mathbf{B}_\Sigma)}_{\mathbf{A}_\Sigma}\cdot\underbrace{\mathbf{Q}\left(\overset{\downarrow}{\mathbf{n}}\cdot\nabla^{\mathrm{T}}\mathbf{Q}\right)}_{\overset{(15.72)}{=}\overset{\downarrow}{\mathbf{n}}\cdot\nabla^{\mathrm{T}}\mathbf{Q}}-\mathbf{n}\cdot\mathbf{B}_\Sigma\cdot\mathbf{Q}\\&+\left(\overset{\downarrow}{\mathbf{C}}_\Sigma\cdot\overset{\downarrow}{\mathbf{n}}\right)\cdot\nabla^{\mathrm{T}}\mathbf{Q}-\boldsymbol{r}_\Sigma\cdot\left(\overset{\downarrow}{\mathbf{B}}_\Sigma\cdot\overset{\downarrow}{\mathbf{n}}\right)\cdot\nabla^{\mathrm{T}}\mathbf{Q}\end{aligned} \tag{15.58}$$

$$\partial_n\mathbf{A}\cdot\mathbf{P}=\mathbf{k}\cdot\mathbf{n}^{\mathrm{T}} \tag{15.59}$$

$$\partial_n\mathbf{A}=\partial_n\mathbf{A}\cdot\mathbf{Q}+\partial_n\mathbf{A}\cdot\mathbf{P} \tag{15.60}$$

* * *

$$\mathbf{A}^{\bullet}_{\Sigma}=\mathbf{0};\quad \partial_n\mathbf{A}^{\bullet}\cdot\mathbf{Q}=\mathbf{0};\quad \partial_n\mathbf{A}^{\bullet}\cdot\mathbf{P}=\mathbf{k}\cdot\mathbf{n}^{\mathrm{T}} \tag{15.61}$$

$$\mathbf{A}^{\bullet}=\operatorname{grad}\mathbf{u};\quad A^{\bullet}_{ik}=\partial_k u_i \tag{15.62}$$

$$\left(A^{\bullet}_{ik}\right)_{\Sigma}=(\partial_k u_i)_{\Sigma}=0 \tag{15.63}$$

$$(\partial_n\mathbf{A}^{\bullet}\cdot\mathbf{Q})_{ik}=n_l\cdot[(Q_{mk}\partial_m)(\partial_l u_i)]_{\Sigma}=0 \tag{15.64}$$

$$(\partial_n\mathbf{A}^{\bullet}\cdot\mathbf{P})_{ik}=n_l\cdot n_m\cdot n_k\cdot(\partial_m\partial_l u_i)_{\Sigma}=\underbrace{\left[(n_l\partial_l)^2u_i\right]_{\Sigma}}_{\partial_n^2 u_i}\cdot n_k=k_i\cdot n_k \tag{15.65}$$

$$(\operatorname{grad}\mathbf{u})_{\Sigma}\overset{(15.63)}{=}\mathbf{0};\quad \partial_n^2\mathbf{u}\overset{(15.65)}{=}\mathbf{k} \tag{15.66}$$

15.4 The Boundary Fields as Functions of Curvilinear Surface Coordinates

In order to perform concrete calculations, one must specify the boundary fields defined on the boundary surface Σ as functions of the curvilinear surface coordinates.

For this purpose, one develops the rows of the matrix fields $\mathbf{B}_{\Sigma}$ and $\mathbf{C}_{\Sigma}$ according to the dual tangent vector basis field (15.8) and the normal vector field (15.5) so that the matrix fields $\mathbf{B}_{\Sigma}$ and $\mathbf{C}_{\Sigma}$ are defined by development coefficients, $\mathbf{b}_1$, $\mathbf{b}_2$, $\mathbf{b}_n$ and, $\mathbf{c}_1$,$\mathbf{c}_2$, $\mathbf{c}_n$ respectively (15.67) and (15.68) which are column matrix fields defined on the boundary surface Σ.

The differential equations, (15.40) and (15.41) by which the matching boundary fields $\mathbf{B}_{\Sigma}$ and $\mathbf{C}_{\Sigma}$ are defined , thereby pass into differential equations for matching column matrix fields $\mathbf{b}_1$, $\mathbf{b}_2$ and $\mathbf{c}_1$, $\mathbf{c}_2$ respectively (15.69) and (15.70). The components of these matrix differential equations consist of independent scalar differential equations which can be solved by known methods. No conditions arise for the matching column matrix fields $\mathbf{b}_n$ and $\mathbf{c}_n$.

The matching $\mathbf{A}$-$\partial_n\mathbf{A}$-boundary fields (15.57)–(15.60) can be expressed by the matching column matrix fields $\mathbf{b}_1$, $\mathbf{b}_2$, $\mathbf{b}_n$; $\mathbf{c}_1$, $\mathbf{c}_2$ $\mathbf{c}_n$ and $\mathbf{k}$ (15.71)–(15.75) where one can choose $\mathbf{b}_n$, $\mathbf{c}_n$ and $\mathbf{k}$ arbitrarily, e.g. set $\mathbf{0}$.[10]

[10] Although one then obtains different $\mathbf{A}$-$\partial_n\mathbf{A}$-boundary fields in each case, the $\mathbf{T}$-solution set (see Fig. 7.2) does not change as a result, which is why each choice of $\mathbf{b}_n$, $\mathbf{c}_n$ and $\mathbf{k}$ leads to the complete general solution.

—

$$\mathbf{B}_\Sigma \overset{\text{def}}{=} \underbrace{\mathbf{b}_1 \cdot \mathbf{f}^{1\mathrm{T}} + \mathbf{b}_2 \cdot \mathbf{f}^{2\mathrm{T}}}_{\mathbf{B}_\Sigma \cdot \mathbf{Q}} + \underbrace{\mathbf{b}_\mathrm{n} \cdot \mathbf{n}^\mathrm{T}}_{\mathbf{B}_\Sigma \cdot \mathbf{P}} \tag{15.67}$$

$$\mathbf{C}_\Sigma \overset{\text{def}}{=} \underbrace{\mathbf{c}_1 \cdot \mathbf{f}^{1\mathrm{T}} + \mathbf{c}_2 \cdot \mathbf{f}^{2\mathrm{T}}}_{\mathbf{C}_\Sigma \cdot \mathbf{Q}} + \underbrace{\mathbf{c}_\mathrm{n} \cdot \mathbf{n}^\mathrm{T}}_{\mathbf{C}_\Sigma \cdot \mathbf{P}} \tag{15.68}$$

* * *

$$|\,\mathbf{f}_1 \times \mathbf{f}_2\,| \cdot (\mathrm{rot}\,\mathbf{B})_\Sigma^\mathrm{T} \cdot \mathbf{n} \overset{(15.38)}{=} \partial_1' \mathbf{b}_2 - \partial_2' \mathbf{b}_1 = |\,\mathbf{f}_1 \times \mathbf{f}_2\,| \cdot \mathbf{t} \tag{15.69}$$

$$|\,\mathbf{f}_1 \times \mathbf{f}_2\,| \cdot (\mathrm{rot}\,\mathbf{C})_\Sigma^\mathrm{T} \cdot \mathbf{n} \overset{(15.38)}{=} \partial_1' \mathbf{c}_2 - \partial_2' \mathbf{c}_1 = |\,\mathbf{f}_1 \times \mathbf{f}_2\,| \cdot \mathbf{r}_\Sigma \times \mathbf{t} \tag{15.70}$$

* * *

$$\mathbf{C}_\Sigma - \mathbf{r}_\Sigma \cdot \mathbf{B}_\Sigma = \underbrace{(\mathbf{c}_1 - \mathbf{r}_\Sigma \times \mathbf{b}_1) \cdot \mathbf{f}^{1\mathrm{T}} + (\mathbf{c}_2 - \mathbf{r}_\Sigma \times \mathbf{b}_2) \cdot \mathbf{f}^{2\mathrm{T}}}_{(\mathbf{C}_\Sigma - \mathbf{r}_\Sigma \cdot \mathbf{B}_\Sigma) \cdot \mathbf{Q}} + \underbrace{(\mathbf{c}_\mathrm{n} - \mathbf{r}_\Sigma \times \mathbf{b}_\mathrm{n}) \cdot \mathbf{n}^\mathrm{T}}_{(\mathbf{C}_\Sigma - \mathbf{r}_\Sigma \cdot \mathbf{B}_\Sigma) \cdot \mathbf{P}} \tag{15.71}$$

$$\begin{aligned} \overset{\downarrow}{\mathbf{n}} \cdot (\nabla^\mathrm{T} \mathbf{Q}) &\overset{(15.25)}{=} \partial_\mu' \mathbf{n} \cdot \mathbf{f}^{\mu\mathrm{T}} \overset{(15.11)}{=} \mathbf{f}^\nu \left(\mathbf{f}_\nu \cdot \partial_\mu' \mathbf{n} \right) \cdot \mathbf{f}^{\mu\mathrm{T}} = -\left(\mathbf{n} \cdot \partial_\mu' \mathbf{f}_\nu \right) \cdot \mathbf{f}^\nu \cdot \mathbf{f}^{\mu\mathrm{T}} \overset{(15.4)}{=} \\ &-\left(\mathbf{n} \cdot \partial_\mu' \partial_\nu' \mathbf{r}_\Sigma \right) \cdot \mathbf{f}^\nu \cdot \mathbf{f}^{\mu\mathrm{T}} = \left[\overset{\downarrow}{\mathbf{n}} \cdot (\nabla^\mathrm{T} \mathbf{Q}) \right]^\mathrm{T} = \mathbf{Q} \cdot \left[\overset{\downarrow}{\mathbf{n}} \cdot (\nabla^\mathrm{T} \mathbf{Q}) \right] \end{aligned} \tag{15.72}$$

$$-\not{\mathbf{n}} \cdot \mathbf{B}_\Sigma \cdot \mathbf{Q} \overset{(15.11),(15.8)}{=} |\mathbf{f}_1 \times \mathbf{f}_2| \left[(\mathbf{b}_1 \mathbf{f}^2) \mathbf{f}^1 \mathbf{f}^{1\mathrm{T}} + (\mathbf{b}_2 \mathbf{f}^2) \mathbf{f}^1 \mathbf{f}^{2\mathrm{T}} - (\mathbf{b}_1 \mathbf{f}^1) \mathbf{f}^2 \mathbf{f}^{1\mathrm{T}} - (\mathbf{b}_2 \mathbf{f}^1) \mathbf{f}^2 \mathbf{f}^{2\mathrm{T}} \right] \tag{15.73}$$

$$\left(\overset{\downarrow}{\mathbf{C}}_\Sigma \cdot \overset{\downarrow}{\mathbf{n}} \right) \cdot \nabla^\mathrm{T} \mathbf{Q} = \overset{\downarrow}{\mathbf{c}}_n \cdot \nabla^\mathrm{T} \mathbf{Q} \overset{(15.25)}{=} \partial_1' \mathbf{c}_n \cdot \mathbf{f}^{1\mathrm{T}} + \partial_2' \mathbf{c}_n \cdot \mathbf{f}^{2\mathrm{T}} \tag{15.74}$$

$$\left(\overset{\downarrow}{\mathbf{B}}_\Sigma \cdot \overset{\downarrow}{\mathbf{n}} \right) \cdot \nabla^\mathrm{T} \mathbf{Q} = \overset{\downarrow}{\mathbf{b}}_n \cdot \nabla^\mathrm{T} \mathbf{Q} \overset{(15.25)}{=} \partial_1' \mathbf{b}_n \cdot \mathbf{f}^{1\mathrm{T}} + \partial_2' \mathbf{b}_n \cdot \mathbf{f}^{2\mathrm{T}} \tag{15.75}$$

Calculation of Special Weightless Stress Tensor Fields

16

Abstract

A "weightless stress tensor field" is presented which generates predetermined boundary stresses. In case the boundary surface on which these boundary stresses are given is not simply connected, another "weightless stress tensor field" is presented, which generates the forces and torques on the continuous boundary surfaces on which no boundary stresses are given, which appear as parameters in the general solution. The superposition of these two stress tensor fields generates both these forces and torques and the specified boundary stresses.

16.1 Calculation of T_*

In this section, a weightless stress tensor field $\mathbf{T}_*$ is computed which generates the specified boundary stresses $\mathbf{t}$ on the boundary surface Σ and which does not generate forces (7.17) and torques (7.18) on the connected boundary surfaces $\Lambda_1, \ldots, \Lambda_n$, on which no boundary stresses are specified.[1]

First of all, it is assumed that the boundary surface Σ on which the boundary stresses $\mathbf{t}$ are given is simply connected or consists of several separate, simply connected pieces. In this case $\mathbf{T}_*$ is any weightless stress tensor field whose boundary stresses on the boundary surface Σ have the given values $\mathbf{t}$. Thus, any $\mathbf{A}_*$-matrix field is sought whose $\mathbf{T}_*$-matrix field satisfies this boundary condition.

First one calculates the $\mathbf{A}$-$\partial_n\mathbf{A}$ – boundary field of this $\mathbf{A}_*$– matrix field defined on the boundary surface Σ. This $\mathbf{A}$-$\partial_n\mathbf{A}$ – boundary-field (15.57)–(15.60) is given by the column-

[1] See Sect. 7.1.

P. Halfar, *Stresses in glaciers*, https://doi.org/10.1007/978-3-662-66024-9_16

matrix-fields $\mathbf{b}_1$, $\mathbf{b}_2$, $\mathbf{b}_n$; $\mathbf{c}_1$, $\mathbf{c}_2$, $\mathbf{c}_n$ and $\mathbf{k}$, (15.71)–(15.75) where one chooses, $\mathbf{b}_n$, $\mathbf{c}_n$ and $\mathbf{k}$ arbitrarily , for example by setting $\mathbf{0}$ and where one chooses for the column-matrix-fields $\mathbf{b}_1$, $\mathbf{b}_2$ and $\mathbf{c}_1$, $\mathbf{c}_2$ any special solution of the differential equations (15.69) and (15.70), which guarantee the given boundary stresses $\mathbf{t}$. A special solution of these differential equations is obtained, for example, by setting $\mathbf{b}_1$ and $\mathbf{c}_1$ to zero and calculating $\mathbf{b}_2$ and $\mathbf{c}_2$ by integration according to the curvilinear surface coordinate x_1'.

For this method a uniform curvilinear coordinate system must be introduced on the boundary surface Σ. If there should be problems with this because of complicated shape of the boundary surface Σ, this boundary surface can be divided by separating lines into several simply shaped surface pieces and the procedure can be carried out separately on each of these surface pieces. However, the column matrix fields $\mathbf{b}_1$, $\mathbf{b}_2$, $\mathbf{c}_1$ and $\mathbf{c}_2$constructed on these surface pieces together do not yet form suitable solutions on the entire boundary surface Σ, since undesirable discontinuities can occur on the separating lines. These discontinuities can be eliminated by changing to other solutions on the individual surface pieces by adding gradient expressions[2] so that continuity is established on the separating lines.

With this $\mathbf{A}$-$\partial_n\mathbf{A}$ – boundary field a suitable $\mathbf{A}_*$ – matrix field can be constructed in the entire glacier area under consideration by starting from its values on the boundary surface Σ and continuing them linearly a small distance according to the normal derivative on rays perpendicular to the boundary surface.[3]The matrix field $\mathbf{A}_*$ constructed in this way in a small vicinity of the boundary surface Σ is then continued sufficiently smoothly over the entire glacier area. This results in an $\mathbf{A}_*$-matrix field whose $\mathbf{T}_*$-matrix field has the given boundary stresses.

If the z-coordinate of the boundary surface Σ can be given as a function $z_0(x, y)$ of the coordinates x and y, it is convenient to introduce the derivative $\partial_z\mathbf{A}_*$ in the z-direction[4] as a boundary condition on the boundary surface Σ instead of the normal derivative of the matrix field $\mathbf{A}_*$. Thus one obtains a suitable $\mathbf{A}_*$-matrix field by linearly continuing its boundary values given on the surface Σ according to this z-derivative.(16.1)

If the boundary surface Σ contains multiple connected parts, any $\mathbf{A}_*$-matrix field is sought whose $\mathbf{T}_*$-matrix field not only has the specified boundary stresses, but also does not generate forces and torques on the connected boundary surfaces $\Lambda_1, \ldots, \Lambda_n$, on which no boundary stresses are specified (7.17) and (7.18).[5]

---

[2] The addition of gradient expressions leads again to solutions of the differential equations (15.69) and (15.70).

[3] These continuations along the rays may only take place in a vicinity of the boundary surface Σ which is so small that different rays cannot intersect there.

[4] This derivative in z-direction can be calculated from derivatives of the matrix field $\mathbf{A}_*$ along the boundary surface Σ and from the normal derivative. One obtains this form of the z-derivative by multiplying the representation (15.23) of the gradient operator from the left by $\mathbf{e}_z^{\mathrm{T}}$.

[5] See Sect. 7.1.

In this case, consider the differential equations (15.69) and (15.70) for the column matrix fields $\mathbf{b}_1$, $\mathbf{b}_2$ and $\mathbf{c}_1$, $\mathbf{c}_2$ on the extended boundary surface $\Sigma \cup \Lambda_1 \cup \ldots \Lambda_n$, where the boundary stresses $\mathbf{t}$ outside of Σ are set to zero by definition. This extended boundary surface contains no multiply connected parts and the differential equations can be solved on this boundary surface using the procedure described above. The weightless stress tensor field $\mathbf{T}_*$ thus constructed then has the boundary stresses $\mathbf{t}$ on surface Σ as well as vanishing boundary stresses and hence vanishing forces and torques on the boundary surfaces $\Lambda_1, \ldots,$ Λ_n.

—

$$\mathbf{A}_*(x, y, z) = [\mathbf{A}_*]_{z=z_0(x,\ y)} + [z - z_0(x,\ y)] \cdot [\partial_z \mathbf{A}_*]_{z=z_0(x,\ y)} \tag{16.1}$$

16.2 Calculation of T_{**}

The weightless stress tensor field $\mathbf{T}_{**}$ occurs in the general solution, if the boundary surface Σ, on which the boundary stresses are given, contains multiple connected components, so that the boundary surface, on which no boundary stresses are given, consists of several separate connected parts $\Lambda_0, \Lambda_1, \ldots \Lambda_n$. On these parts, with the exception of Λ_0, shall $\mathbf{T}_{**}$ generate forces $\mathbf{F}_1, \ldots \mathbf{F}_n$ and torques $\mathbf{G}_1, \ldots \mathbf{G}_n$, which represent the free parameters of the general solution. At the same time the boundary stresses of $\mathbf{T}_{**}$ everywhere on the boundary surface Σ shall disappear . Therefore a matrix field $\mathbf{A}_{**}$ is to be constructed on the considered glacier area, whose $\mathbf{T}_{**}$-field fulfills these boundary conditions.

This task can be solved with the help of matrix fields $\mathbf{B}_{**}$ and $\mathbf{C}_{**}$ which descend from the matrix field $\mathbf{A}_{**}$. So the task is to find matrix fields $\mathbf{B}_{**}$ and $\mathbf{C}_{**}$ which satisfy the descent condition (16.2), which lead to vanishing boundary stresses on the boundary surface Σ (16.3) and which generate forces $\mathbf{F}_1, \ldots \mathbf{F}_n$ (16.4) and torques $\mathbf{G}_1, \ldots \mathbf{G}_n$ (16.5) on the surfaces $\Lambda_1, \ldots \Lambda_n$.

First, the simpler tasks are considered, in which in each case only on one surface $\Lambda_\mu(\mu = 1, \ldots n)$ among the surfaces $\Lambda_1, \ldots \Lambda_n$, a force $\mathbf{F}_\mu$ and a torque $\mathbf{G}_\mu$ occur, while on the other surfaces the forces and torques disappear. These tasks are solved using vector fields $\mathbf{w}_\mu$, which have the following properties[6]: The path integral of $\mathbf{w}_\mu$ over the oriented edge of the surface Λ_μ has value one and the rotation of the vector field $\mathbf{w}_\mu$ vanishes on a spatial neighborhood Ω'_μ of the boundary surfaces $\Sigma;\Lambda_1, \ldots, \Lambda_n$-except Λ_μ-, which is why, by Stokes' theorem, the path integrals of $\mathbf{w}_\mu$ over the oriented edges of the surface $\Lambda_1, \ldots \Lambda_n$ vanish except for Λ_μ. (16.6)–(16.8) These vector fields $\mathbf{w}_\mu$ are used to construct $\mathbf{B}_\mu$ and $\mathbf{C}_\mu$ tensor fields on the spatial neighborhoods Ω'_μ, which are solutions of the above simpler

[6] Such vector fields $\mathbf{w}_\mu$ are specified below.

tasks and which derive from tensor fields $\mathbf{A}_\mu$ (16.9)–(16.16). The corresponding weightless stress tensor fields $\mathbf{T}_\mu$ disappear on the spatial neighborhood Ω'_μ (16.13) and they generate on the surface Λ_μ the force $\mathbf{F}_\mu$ and torque $\mathbf{G}_\mu$ (16.14) and (16.15) and due to the balance conditions they generate on the surface Λ_0 the force $-\mathbf{F}_\mu$ and torque $-\mathbf{G}_\mu$.

The tensor fields $\mathbf{A}_\mu$ (16.16), on the other hand, are defined in Ω'_μ each case not only on, but wherever the vector field $\mathbf{w}_\mu$ is defined.[7] Their sum $\mathbf{A}_{**}$ and the tensor fields descending from it $\mathbf{B}_{**}$, $\mathbf{C}_{**}$ and $\mathbf{T}_{**}$ (16.17)–(16.22) form a solution of the posed problem, (16.2)–(16.5) which depends linearly on the forces $\mathbf{F}_\mu$ and the torques $\mathbf{G}_\mu$ on the surfaces Λ_μ ($\mu = 1, \ldots, n$) of unknown boundary stresses.

The $\mathbf{w}_\mu$ vector fields are still missing. It is assumed that there is an oriented axis of rotation to the respective boundary surface Λ_μ, which enters the considered glacier area Ω through the boundary surface Λ_0, leaves it through the boundary surface Λ_μ and does not intersect the closed boundary $\partial\Omega$ otherwise . The angular coordinate ϕ[8] defined with respect to this axis of rotation is an ambiguous function. Its gradient field (16.27) divided 2π by, on the other hand, is unambiguous and already has the properties that the vector field $\mathbf{w}_\mu$ is also supposed to have (16.6)–(16.8). This gradient field does have an undesirable singularity on the axis of rotation, but this can be eliminated by multiplying it by an interpolation function $\chi(R)$ (16.29) that depends on the axial distance R without damaging the required properties. To this end, one chooses this interpolation function $\chi(R)$ to be equal to 1 everywhere except in a small cylindrical neighborhood of the axis of rotation, where this function approaches zero sufficiently rapidly as the axis of rotation is approached. Thus there is a suitable vector field $\mathbf{w}_\mu$ (16.29) defined in the whole space.

If there is no such axis of rotation, then the glacier area under consideration is mathematically mapped in such a way that there is a suitable axis of rotation for the image area. If one transfers the corresponding angular coordinate divided by 2π from the image domain to the considered glacier domain by the inverse mapping, one obtains an ambiguous function, its gradient field has the desired properties and the singularity[9] can be eliminated similarly as above.

[7] The vector fields $\mathbf{w}_\mu$ constructed below are defined in the whole space. Thus, each tensor field $\mathbf{A}_\mu$ is also defined in the whole space. It generates by the weightless stress tensor field descending from it on the surface Λ_μ a force $\mathbf{F}_\mu$ and a torque $\mathbf{G}_\mu$, on the surface Λ_0 a force $-\mathbf{F}_\mu$ and a torque $-\mathbf{G}_\mu$, on the remaining part of the closed boundary $\partial\Omega$ thus on the other surfaces Λ_ν as well as on the surface Σ no boundary stresses and thus also no forces and torques.

[8] The angular coordinate ϕ and other quantities in equations (16.23)–(16.31) are different for different surfaces Λ_μ. So one has to add the subscript "μ", which is omitted from the formulas so that they do not become too cumbersome. It denote $\mathbf{a}$ the unit vector in the direction of the axis, $\hat{\mathbf{r}}$ the vector from the origin of coordinates to a point on the axis of rotation, $\mathbf{R}$ the vector from the axis and perpendicular to it to the point under consideration, the $\mathbf{r}$ vector from the origin of coordinates to the point under consideration, and R the distance of the point under consideration from the axis of rotation.

[9] The singularity occurs on the line into which the axis of rotation passes by inversion.

—

$$(\text{rot}\mathbf{C}_{**})^{\text{T}} = \not{\mathbf{r}} \cdot (\text{rot}\mathbf{B}_{**})^{\text{T}} \tag{16.2}$$

$$[\mathbf{T}_{**} \cdot \mathbf{n}]_{\Sigma} \overset{\text{id.}}{=} \left[(\text{rot}\mathbf{B}_{**})^{\text{T}} \cdot \mathbf{n}\right]_{\Sigma} = \mathbf{0} \tag{16.3}$$

$$\int_{\Lambda_\nu} \mathbf{T}_{**}\mathbf{n} \cdot \text{d}A \overset{\text{id.}}{=} \oint_{\partial\Lambda_\nu} \mathbf{B}_{**} \cdot \mathbf{dr} = \mathbf{F}_\nu; \quad \nu = 1, \ldots, n \tag{16.4}$$

$$\int_{\Lambda_\nu} \mathbf{r} \times \mathbf{T}_{**}\mathbf{n} \cdot \text{d}A \overset{\text{id.}}{=} \oint_{\partial\Lambda_\nu} \mathbf{C}_{**} \cdot \mathbf{dr} = \mathbf{G}_\nu; \quad \nu = 1, \ldots, n \tag{16.5}$$

* * *

$$\mu,\nu = 1, \ldots, n : \quad \oint_{\partial\Lambda_\mu} w_\mu \cdot \mathbf{dr} \overset{\text{prec.}}{=} 1 \tag{16.6}$$

$$\nabla \times \mathbf{w}_\mu \overset{prec.}{=} \mathbf{0}; \quad \mathbf{r} \in \Omega'_\mu \supset \partial\Omega - \{\Lambda_0 \cup \Lambda_\mu\} \tag{16.7}$$

$$\oint_{\partial\Lambda_\nu} \mathbf{w}_\mu \cdot \mathbf{dr} = \delta_{\mu\nu} \tag{16.8}$$

* * *

$$\mu,\nu = 1, \ldots, n : \qquad \mathbf{B}_\mu(\mathbf{r}) \overset{\text{def}}{=} \mathbf{F}_\mu \mathbf{w}_\mu^{\text{T}}(\mathbf{r}); \quad \mathbf{r} \in \Omega'_\mu; \text{no sum} \tag{16.9}$$

$$\mathbf{C}_\mu(\mathbf{r}) \overset{\text{def}}{=} \mathbf{G}_\mu \mathbf{w}_\mu^{\text{T}}(\mathbf{r}); \quad \mathbf{r} \in \Omega'_\mu; \text{no sum} \tag{16.10}$$

$$\text{rot}\,\mathbf{B}_\mu = \text{rot}\,\mathbf{C}_\mu \overset{(16.7)}{=} 0; \quad \mathbf{r} \in \Omega'_\mu \tag{16.11}$$

$$(\text{rot}\,\mathbf{C}_\mu)^{\text{T}} = \not{\mathbf{r}} \cdot (\text{rot}\,\mathbf{B}_\mu)^{\text{T}}; \quad \mathbf{r} \in \Omega'_\mu \tag{16.12}$$

$$\mathbf{T}_\mu = (\text{rot}\,\mathbf{B}_\mu)^{\text{T}} = \mathbf{0}; \quad \mathbf{r} \in \Omega'_\mu \tag{16.13}$$

$$\oint_{\partial\Lambda_\nu} \mathbf{B}_\mu \cdot \mathbf{dr} \overset{(16.8)}{=} \delta_{\mu\nu} \cdot \mathbf{F}_\mu; \quad \text{no sum} \tag{16.14}$$

$$\oint_{\partial\Lambda_\nu} \mathbf{C}_\mu \cdot \mathrm{d}\mathbf{r} \overset{(16.8)}{=} \delta_{\mu\nu} \cdot \mathbf{G}_\mu; \quad \text{no sum} \tag{16.15}$$

$$\mathbf{A}_\mu(\mathbf{r}) = \underbrace{\mathbf{G}_\mu \cdot \mathbf{w}_\mu^{\mathrm{T}} - \mathbf{r} \cdot \mathbf{F}_\mu \cdot \mathbf{w}_\mu^{\mathrm{T}}}_{= \mathbf{C}_\mu - \mathbf{r} \cdot \mathbf{B}_\mu \text{ in } \Omega'_\mu}; \quad \text{no sum} \tag{16.16}$$

* * *

$$\mathbf{A}_{**} = (\mathbf{G}_\mu + \mathbf{F}_\mu \cdot \mathbf{r}) \cdot \mathbf{w}_\mu^{\mathrm{T}} = (\mathbf{G}_\mu - \mathbf{r} \cdot \mathbf{F}_\mu) \cdot \mathbf{w}_\mu^{\mathrm{T}} \tag{16.17}$$

$$\mathrm{rot}\,\mathbf{A}_{**} = \mathbf{F}_\mu \cdot \mathbf{w}_\mu^{\mathrm{T}} + (\nabla \times \mathbf{w}_\mu)\left(\mathbf{G}_\mu^{\mathrm{T}} - \mathbf{r}^{\mathrm{T}} \cdot \mathbf{F}_\mu\right) - \left(\mathbf{w}_\mu^{\mathrm{T}} \cdot \mathbf{F}_\mu\right) \cdot 1 \tag{16.18}$$

$$\begin{aligned} \mathrm{rot\,rot}\,\mathbf{A}_{**} &= (\nabla \times \mathbf{w}_\mu) \cdot \mathbf{F}_\mu^{\mathrm{T}} + 2\mathbf{F}_\mu \cdot (\nabla \times \mathbf{w}_\mu)^{\mathrm{T}} \\ &+ \left(\mathbf{F}_\mu \mathbf{r}^{\mathrm{T}} - \mathbf{r}\mathbf{F}_\mu^{\mathrm{T}} - \mathbf{G}_\mu\right) \cdot \nabla \cdot (\nabla \times \mathbf{w}_\mu)^{\mathrm{T}} - \nabla\left(\mathbf{F}_\mu^{\mathrm{T}} \cdot \mathbf{w}_\mu\right) \end{aligned} \tag{16.19}$$

$$\begin{aligned} \mathbf{B}_{**} &= \mathrm{rot}\,\mathbf{A}_{**} - 1 \cdot \frac{1}{2} \cdot \mathrm{trace}(\mathrm{rot}\mathbf{A}_{**}) = \mathbf{F}_\mu \cdot \mathbf{w}_\mu^{\mathrm{T}} \\ &+ (\nabla \times \mathbf{w}_\mu)\left(\mathbf{G}_\mu^{\mathrm{T}} - \mathbf{r}^{\mathrm{T}} \cdot \mathbf{F}_\mu\right) - 1 \cdot \frac{1}{2} \cdot \left[\left(\mathbf{G}_\mu^{\mathrm{T}} - \mathbf{r}^{\mathrm{T}} \cdot \mathbf{F}_\mu\right) \cdot (\nabla \times \mathbf{w}_\mu)\right] \end{aligned} \tag{16.20}$$

$$\begin{aligned} \mathbf{C}_{**} &= \mathbf{A}_{**} + \mathbf{r} \cdot \mathbf{B}_{**} = \mathbf{G}_\mu \cdot \mathbf{w}_\mu^{\mathrm{T}} \\ &+ \mathbf{r} \cdot \left\{(\nabla \times \mathbf{w}_\mu)\left(\mathbf{G}_\mu^{T} - \mathbf{r}^{\mathrm{T}} \cdot \mathbf{F}_\mu\right) - 1 \cdot \frac{1}{2} \cdot \left[\left(\mathbf{G}_\mu^{\mathrm{T}} - \mathbf{r}^{\mathrm{T}} \cdot \mathbf{F}_\mu\right) \cdot (\nabla \times \mathbf{w}_\mu)\right]\right\} \end{aligned} \tag{16.21}$$

$$\mathbf{T}_{**} = \frac{1}{2}\left[\mathrm{rot\,rot}\,\mathbf{A}_{**} + (\mathrm{rot\,rot}\mathbf{A}_{**})^{\mathrm{T}}\right] = (\mathrm{rot\,rot}\,\mathbf{A}_{**})_+ \tag{16.22}$$

* * *

$$|\,\mathbf{a}\,| = 1 \tag{16.23}$$

$$\mathbf{R} = \mathbf{r} - \hat{\mathbf{r}} - [\mathbf{a} \cdot (\mathbf{r} - \hat{\mathbf{r}})] \cdot \mathbf{a} \tag{16.24}$$

$$R^2 = |\mathbf{R}|^2 = (\mathbf{r} - \hat{\mathbf{r}})^2 - [\mathbf{a} \cdot (\mathbf{r} - \hat{\mathbf{r}})]^2 \tag{16.25}$$

$$\nabla R = \frac{\mathbf{R}}{R} \tag{16.26}$$

$$\frac{1}{2\pi} \cdot \nabla\phi = \frac{1}{2\pi R} \cdot \mathbf{a} \times \frac{\mathbf{R}}{R} \tag{16.27}$$

$$\nabla \times \frac{1}{2\pi} \cdot \nabla\phi = \mathbf{0}; R \neq 0 \tag{16.28}$$

$$\mathbf{w}_\mu = \frac{\chi(R) \cdot \nabla\phi}{2\pi} \tag{16.29}$$

$$\nabla \times \mathbf{w}_\mu = \frac{\chi'(R)}{2\pi R} \cdot \mathbf{a} \tag{16.30}$$

$$\nabla \cdot \left(\nabla \times \mathbf{w}_\mu\right)^{\mathrm{T}} = \frac{\left[R \cdot \chi'(R) - 2\chi(R)\right]'}{2\pi R^3} \cdot \mathbf{R} \cdot \mathbf{a}^{\mathrm{T}} \tag{16.31}$$

The General Solution Expressed by Three Independent Stress Components

17

Abstract

Eight models “a” to “h” are listed in detail, which describe the general solution of the balance and boundary conditions by three selected components of the stress tensor or the deviatoric stress tensor. These three selected components can each be taken as arbitrary functions within the framework of the general solution.

The eight combinations “a” to “h” of three independent stress components of $\mathbf{T}_0$ in Table 8.1 in Sect. 8.2.2 are examples of the eight following combination types, where the number of combinations of a type is given in brackets in each case:

(a) Three diagonals (1)
(b) Two diagonals and one in the intersection field of the two (3)
(c) Two diagonals and one not in the intersection field (6)
(d) One diagonal (6)
(e) No diagonal (1)
(f) Two deviatoric diagonals and one deviatoric in the intersection field of the two (3)
(g) Two deviatoric diagonals and one deviatoric not in the intersection field (6)
(h) One deviatoric diagonal (6)

All combinations of a type emerge from each other by swapping the Cartesian location coordinates.

In the cases “a” to “e” there are a total of 17 combinations. These are the 20 combinatorial possibilities to select three out of six matrix elements, minus the three “forbidden” cases, in which the three matrix elements are each in a row or column, because these three

P. Halfar, *Stresses in glaciers*, https://doi.org/10.1007/978-3-662-66024-9_17

matrix elements are not independent of each other due to vanishing divergence of the rows or columns.

In the cases “e” to “h” there are a total of 16 combinations. These are again the 20 combinatorial possibilities to choose three out of six matrix elements – this time of the deviatoric tensor – this time minus the 4 “forbidden” following cases: This is once the combination of the 3 diagonal deviatoric matrix elements, since their sum vanishes and these matrix elements are therefore not independent of each other, and these are the three combinations in which the three deviatoric matrix elements are each in a row or column, since in these cases elliptic differential equations occur which have no unique solution under the present conditions.

In the following, for the combinations “a” to “h” of independent stress components given in Table 8.1, the weightless solutions $\mathbf{T}_0$ with given independent stress components and the solutions $\mathbf{S}_* = \mathbf{S}_a$ up to $\mathbf{S}_* = \mathbf{S}_h$ of the balance and boundary conditions with vanishing independent stress components are calculated.[1] The following elements are decisive for the respective solution and are specified:

- The occurring integral operators
- The convex integration cones[2] of the integral operators
- The generating cone vectors of the integration cones
- The convex model cone generated by the integration cones
- The $\mathbf{A}_0$-normalization and its normalization directions in the mother model[3]
- The differential equations for the three independent matrix elements of the normalized matrix field $\mathbf{A}_0$ from the mother model
- The matrix field $\mathbf{A}_0$ of the mother model
- The matrix field $\mathbf{B}_0 \overset{(6.8)}{=} \mathrm{rot}\mathbf{A}_0 - \frac{1}{2} \cdot \mathrm{trace}(\mathrm{rot}\mathbf{A}_0) \cdot \mathbf{1}$
- The weightless solution $\mathbf{T}_0 \overset{(6.9)}{=} (\mathrm{rot}\,\mathbf{B}_0)^{\mathrm{T}} \overset{\mathrm{id.}}{=} \mathrm{rot}\,\mathrm{rot}\,\mathbf{A}_{0+} \overset{\mathrm{id.}}{=} [\mathrm{rot}\,\mathrm{rot}\,\mathbf{A}_0]_+$ in dependence of its arbitrarily specified independent stress components
- The solution $\mathbf{S}_*$ ($=\mathbf{S}_a, \ldots, \mathbf{S}_h$) with vanishing independent stress components
- The boundary values of $\mathbf{T}_0$ at the free surface Σ
- The boundary values of $\mathbf{S}_*$ at the free surface Σ (these disappear)

The model “h” with three independent deviatoric *xy*-, *yz*- and *xx*-components is a special case, because in this case two solutions with different model and integration cones are possible. Each of the formulas given in this model “h”[4] is valid for both solutions, but

---

[1] See Sects. 8.2 and 3.4.

[2] The cone designations are defined in Sect. 8.2.2.

[3] Since the free surface Σ should be transverse and synchronous to the model cone (see Sect. 8.2.3), the respective $\mathbf{A}_0$ normalization is also permissible, since the normalization directions are then also transverse to the free surface Σ (see (7.26)).

[4] See Sect. 17.8.

defines different calculation rules, since the one-dimensional integration cone of the integral operator $\left(\partial_x - \sqrt{2}\partial_z\right)^{-1}$ for one of the two solutions is generated by the vector $-\mathbf{e}_x + \sqrt{2}\mathbf{e}_z$, and for the other solution by the opposite vector $\mathbf{e}_x - \sqrt{2}\mathbf{e}_z$.[5]

$$\Sigma: \quad z = z_0(x, y) \text{ or } y = y_0(x, z) \tag{17.1}$$

$$[\cdot]_{z_0} \overset{\text{def.}}{=} [\cdot]_{z = z_0(x, y)} \tag{17.2}$$

$$[\cdot]_{y_0} \overset{\text{def.}}{=} [\cdot]_{y = y_0(x, z)} \tag{17.3}$$

17.1 (a) Independent *xx*-, *yy*-, *zz*-Components

Integral operators	∂_x^{-1}	∂_y^{-1}	∂_z^{-1}
Integration cones	K_x	K_y	K_z
Generative cone vectors	$\mathbf{e}_x$	$\mathbf{e}_y$	$\mathbf{e}_z$
Model cone	K_{xyz}		
A_0 normalization	$xx - yy - xy$		
Normalization directions	z		

$$\underbrace{\begin{bmatrix} 0 & \partial_z^2 & 0 \\ \partial_z^2 & 0 & 0 \\ \partial_y^2 & \partial_x^2 & -2\partial_x\partial_y \end{bmatrix}}_{\mathcal{L}} \underbrace{\begin{bmatrix} A_{0xx} \\ A_{0yy} \\ A_{0xy} \end{bmatrix}}_{f} = \underbrace{\begin{bmatrix} T_{0xx} \\ T_{0yy} \\ T_{0zz} \end{bmatrix}}_{q} \tag{17.4}$$

$$\mathcal{L}^{-1} = \frac{1}{2} \cdot \begin{bmatrix} 0 & 2 & 0 \\ 2 & 0 & 0 \\ \partial_x\partial_y^{-1} & \partial_x^{-1}\partial_y & -\partial_x^{-1}\partial_y^{-1}\partial_z^2 \end{bmatrix} \cdot \partial_z^{-2} \tag{17.5}$$

[5] See footnote 11, Sect. 3.2.

$$\mathbf{A}_0=\mathbf{A}_0^{\mathrm{T}}=\frac{1}{2}\begin{bmatrix}0 & \partial_x\partial_y^{-1}\partial_z^{-2} & 0\\ * & 2\partial_z^{-2} & 0\\ 0 & 0 & 0\end{bmatrix}T_{0xx}+\frac{1}{2}\begin{bmatrix}2\partial_z^{-2} & \partial_x^{-1}\partial_y\partial_z^{-2} & 0\\ * & 0 & 0\\ 0 & 0 & 0\end{bmatrix}T_{0yy}+\frac{1}{2}$$
$$\times\begin{bmatrix}0 & -\partial_x^{-1}\partial_y^{-1} & 0\\ * & 0 & 0\\ 0 & 0 & 0\end{bmatrix}T_{0zz} \tag{17.6}$$

$$\mathbf{B}_0=\frac{1}{2}\begin{bmatrix}-\partial_x\partial_y^{-1}\partial_z^{-1} & -2\partial_z^{-1} & 0\\ 0 & \partial_x\partial_y^{-1}\partial_z^{-1} & 0\\ \partial_x^2\partial_y^{-1}\partial_z^{-2} & \partial_x\partial_z^{-2} & 0\end{bmatrix}T_{0xx}+\frac{1}{2}\begin{bmatrix}-\partial_x^{-1}\partial_y\partial_z^{-1} & 0 & 0\\ 2\partial_z^{-1} & \partial_x^{-1}\partial_y\partial_z^{-1} & 0\\ -\partial_y\partial_z^{-2} & -\partial_x^{-1}\partial_y^2\partial_z^{-2} & 0\end{bmatrix}T_{0yy}$$
$$+\frac{1}{2}\begin{bmatrix}\partial_z\partial_x^{-1}\partial_y^{-1} & 0 & 0\\ 0 & -\partial_z\partial_x^{-1}\partial_y^{-1} & 0\\ -\partial_y^{-1} & \partial_x^{-1} & 0\end{bmatrix}T_{0zz} \tag{17.7}$$

$$\mathbf{T}_0=\mathbf{T}_0^{\mathrm{T}}=\frac{1}{2}\begin{bmatrix}2 & -\partial_x\partial_y^{-1} & -\partial_x\partial_z^{-1}\\ * & 0 & \partial_x^2\partial_y^{-1}\partial_z^{-1}\\ * & * & 0\end{bmatrix}T_{0xx}+\frac{1}{2}\begin{bmatrix}0 & -\partial_y\partial_x^{-1} & \partial_y^2\partial_x^{-1}\partial_z^{-1}\\ * & 2 & -\partial_y\partial_z^{-1}\\ * & * & 0\end{bmatrix}T_{0yy}+\frac{1}{2}$$
$$\times\begin{bmatrix}0 & \partial_z^2\partial_x^{-1}\partial_y^{-1} & -\partial_z\partial_x^{-1}\\ * & 0 & -\partial_z\partial_y^{-1}\\ * & * & 2\end{bmatrix}T_{0zz} \tag{17.8}$$

$$\mathbf{S}_*=\mathbf{S}_{\mathrm{a}}=\mathbf{S}_{\mathrm{a}}^{\mathrm{T}}=\frac{1}{2}\begin{bmatrix}0 & (-g_x\partial_x-g_y\partial_y+g_z\partial_z)\cdot\partial_x^{-1}\partial_y^{-1} & (-g_x\partial_x+g_y\partial_y-g_z\partial_z)\cdot\partial_x^{-1}\partial_z^{-1}\\ * & 0 & (g_x\partial_x-g_y\partial_y-g_z\partial_z)\cdot\partial_y^{-1}\partial_z^{-1}\\ * & * & 0\end{bmatrix}\rho \tag{17.9}$$

$$[\mathbf{T}_0]_{z0}=\left[\mathbf{T}_0^{\mathrm{T}}\right]_{z0}=\frac{1}{2}\begin{bmatrix}2 & -\partial_x z_0/\partial_y z_0 & \partial_x z_0\\ * & 0 & -(\partial_x z_0)^2/\partial_y z_0\\ * & * & 0\end{bmatrix}[T_{0xx}]_{z0}+\frac{1}{2}$$
$$\times\begin{bmatrix}0 & -\partial_y z_0/\partial_x z_0 & -(\partial_y z_0)^2/\partial_x z_0\\ * & 2 & \partial_y z_0\\ * & * & 0\end{bmatrix}[T_{0yy}]_{z0}+\frac{1}{2}$$
$$\times\begin{bmatrix}0 & 1/(\partial_x z_0\cdot\partial_y z_0) & 1/\partial_x z_0\\ * & 0 & 1/\partial_y z_0\\ * & * & 2\end{bmatrix}[T_{0zz}]_{z0} \tag{17.10}$$

$$[\mathbf{S}_*]_{z0}=[\mathbf{S}_a]_{z0}=\mathbf{0} \tag{17.11}$$

17.2 (b) Independent *xx*-, *yy*-, *xy*-Components

Integral operators	∂_z^{-1}
Integration cones	K_z
Generative cone vectors	$\mathbf{e}_z$
Model cone	K_z
A_0 normalization	$xx - yy - xy$
Normalization directions	z

$$\underbrace{\begin{bmatrix} \partial_z^2 & 0 & 0 \\ 0 & \partial_z^2 & 0 \\ 0 & 0 & -\partial_z^2 \end{bmatrix}}_{\mathcal{L}} \underbrace{\begin{bmatrix} A_{0xx} \\ A_{0yy} \\ A_{0xy} \end{bmatrix}}_{f} = \underbrace{\begin{bmatrix} T_{0yy} \\ T_{0xx} \\ T_{0xy} \end{bmatrix}}_{q} \tag{17.12}$$

$$\mathcal{L}^{-1} = \begin{bmatrix} 1 & 0 & 0 \\ 0 & 1 & 0 \\ 0 & 0 & -1 \end{bmatrix} \cdot \partial_z^{-2} \tag{17.13}$$

$$\mathbf{A}_0 = \mathbf{A}_0^{\mathrm{T}} = \begin{bmatrix} 0 & 0 & 0 \\ 0 & \partial_z^{-2} & 0 \\ 0 & 0 & 0 \end{bmatrix} T_{0xx} + \begin{bmatrix} \partial_z^{-2} & 0 & 0 \\ 0 & 0 & 0 \\ 0 & 0 & 0 \end{bmatrix} T_{0yy} + \begin{bmatrix} 0 & -\partial_z^{-2} & 0 \\ * & 0 & 0 \\ 0 & 0 & 0 \end{bmatrix} T_{0xy} \tag{17.14}$$

$$\mathbf{B}_0 = \begin{bmatrix} 0 & -\partial_z^{-1} & 0 \\ 0 & 0 & 0 \\ 0 & \partial_x\partial_z^{-2} & 0 \end{bmatrix} T_{0xx} + \begin{bmatrix} 0 & 0 & 0 \\ \partial_z^{-1} & 0 & 0 \\ -\partial_y\partial_z^{-2} & 0 & 0 \end{bmatrix} T_{0yy} + \begin{bmatrix} \partial_z^{-1} & 0 & 0 \\ 0 & -\partial_z^{-1} & 0 \\ -\partial_x\partial_z^{-2} & \partial_y\partial_z^{-2} & 0 \end{bmatrix} T_{0xy} \tag{17.15}$$

$$\begin{aligned} \mathbf{T}_0 = \mathbf{T}_0^{\mathrm{T}} = {} & \begin{bmatrix} 1 & 0 & -\partial_x\partial_z^{-1} \\ 0 & 0 & 0 \\ * & 0 & \partial_x^2\partial_z^{-2} \end{bmatrix} T_{0xx} + \begin{bmatrix} 0 & 0 & 0 \\ 0 & 1 & -\partial_y\partial_z^{-1} \\ 0 & * & \partial_y^2\partial_z^{-2} \end{bmatrix} T_{0yy} \\ & + \begin{bmatrix} 0 & 1 & -\partial_y\partial_z^{-1} \\ 1 & 0 & -\partial_x\partial_z^{-1} \\ * & * & 2\partial_x\partial_y\partial_z^{-2} \end{bmatrix} T_{0xy} \end{aligned} \tag{17.16}$$

$$\mathbf{S}_* = \mathbf{S}_b = \mathbf{S}_b^{\mathrm{T}} = \begin{bmatrix} 0 & 0 & -g_x\partial_z^{-1} \\ 0 & 0 & -g_y\partial_z^{-1} \\ * & * & (g_x\partial_x + g_y\partial_y - g_z\partial_z)\cdot\partial_z^{-2} \end{bmatrix} \rho \tag{17.17}$$

$$[\mathbf{T}_0]_{z_0} = \left[\mathbf{T}_0^{\mathrm{T}}\right]_{z_0} = \begin{bmatrix} 1 & 0 & \partial_x z_0 \\ 0 & 0 & 0 \\ * & 0 & (\partial_x z_0)^2 \end{bmatrix} [T_{0xx}]_{z_0} + \begin{bmatrix} 0 & 0 & 0 \\ 0 & 1 & \partial_y z_0 \\ * & * & (\partial_y z_0)^2 \end{bmatrix} \left[T_{0yy}\right]_{z_0}$$
$$+ \begin{bmatrix} 0 & 1 & \partial_y z_0 \\ 1 & 0 & \partial_x z_0 \\ * & * & 2\partial_x z_0 \cdot \partial_y z_0 \end{bmatrix} \left[T_{0xy}\right]_{z_0} \tag{17.18}$$

$$[\mathbf{S}_*]_{z_0} = [\mathbf{S}_b]_{z_0} = \mathbf{0} \tag{17.19}$$

17.3 (c) Independent *xx*-, *yy*-, *xz*-Components

Integral operators	∂_y^{-1}	∂_z^{-1}
Integration cones	K_y	K_z
Generative cone vectors	$\mathbf{e}_y$	$\mathbf{e}_z$
Model cone	K_{yz}	
A_0 normalization	$xx - yy - xy$	
Normalization directions	z	

$$\underbrace{\begin{bmatrix} \partial_y\partial_z & 0 & -\partial_x\partial_z \\ 0 & \partial_z^2 & 0 \\ 0 & 0 & \partial_z^2 \end{bmatrix}}_{\mathcal{L}} \underbrace{\begin{bmatrix} A_{0xy} \\ A_{0xx} \\ A_{0yy} \end{bmatrix}}_{f} = \underbrace{\begin{bmatrix} T_{0xz} \\ T_{0yy} \\ T_{0xx} \end{bmatrix}}_{q} \tag{17.20}$$

$$\mathcal{L}^{-1} = \begin{bmatrix} \partial_y^{-1}\partial_z & 0 & \partial_x\partial_y^{-1} \\ 0 & 1 & 0 \\ 0 & 0 & 1 \end{bmatrix} \cdot \partial_z^{-2} \tag{17.21}$$

$$\mathbf{A}_0 = \mathbf{A}_0^{\mathrm{T}} = \begin{bmatrix} 0 & \partial_x\partial_y^{-1}\partial_z^{-2} & 0 \\ * & \partial_z^{-2} & 0 \\ 0 & 0 & 0 \end{bmatrix} T_{0xx} + \begin{bmatrix} \partial_z^{-2} & 0 & 0 \\ 0 & 0 & 0 \\ 0 & 0 & 0 \end{bmatrix} T_{0yy} + \begin{bmatrix} 0 & \partial_y^{-1}\partial_z^{-1} & 0 \\ * & 0 & 0 \\ 0 & 0 & 0 \end{bmatrix} T_{0xz} \tag{17.22}$$

$$\mathbf{B}_0 = \begin{bmatrix} -\partial_x\partial_y^{-1}\partial_z^{-1} & -\partial_z^{-1} & 0 \\ 0 & \partial_x\partial_y^{-1}\partial_z^{-1} & 0 \\ \partial_x^2\partial_y^{-1}\partial_z^{-2} & 0 & 0 \end{bmatrix} T_{0xx} + \begin{bmatrix} 0 & 0 & 0 \\ \partial_z^{-1} & 0 & 0 \\ -\partial_y\partial_z^{-2} & 0 & 0 \end{bmatrix} T_{0yy}$$
$$+ \begin{bmatrix} -\partial_y^{-1} & 0 & 0 \\ 0 & \partial_y^{-1} & 0 \\ \partial_x\partial_y^{-1}\partial_z^{-1} & -\partial_z^{-1} & 0 \end{bmatrix} T_{0xz} \tag{17.23}$$

$$\mathbf{T}_0=\mathbf{T}_0^{\mathrm{T}}=\begin{bmatrix}1 & -\partial_x\partial_y^{-1} & 0\\ * & 0 & \partial_x^2\partial_y^{-1}\partial_z^{-1}\\ 0 & * & -\partial_x^2\partial_z^{-2}\end{bmatrix}T_{0xx}+\begin{bmatrix}0 & 0 & 0\\ 0 & 1 & -\partial_y\partial_z^{-1}\\ 0 & * & \partial_y^2\partial_z^{-2}\end{bmatrix}T_{0yy}$$
$$+\begin{bmatrix}0 & -\partial_z\partial_y^{-1} & 1\\ * & 0 & \partial_x\partial_y^{-1}\\ 1 & * & -2\partial_x\partial_z^{-1}\end{bmatrix}T_{0xz} \tag{17.24}$$

$$\mathbf{S}_*=\mathbf{S}_c=\mathbf{S}_c^{\mathrm{T}}=\begin{bmatrix}0 & -g_x\partial_y^{-1} & 0\\ * & 0 & \left(g_x\partial_x-g_y\partial_y\right)\cdot\partial_y^{-1}\partial_z^{-1}\\ 0 & * & \left(-g_x\partial_x+g_y\partial_y-g_z\partial_z\right)\cdot\partial_z^{-2}\end{bmatrix}\rho \tag{17.25}$$

$$[\mathbf{T}_0]_{z0}=\left[\mathbf{T}_0^{\mathrm{T}}\right]_{z0}=\begin{bmatrix}1 & -\partial_x z_0/\partial_y z_0 & 0\\ * & 0 & -(\partial_x z_0)^2/\partial_y z_0\\ 0 & * & -(\partial_x z_0)^2\end{bmatrix}[T_{0xx}]_{z0}$$
$$+\begin{bmatrix}0 & 0 & 0\\ 0 & 1 & \partial_y z_0\\ 0 & * & \left(\partial_y z_0\right)^2\end{bmatrix}\left[T_{0yy}\right]_{z0}+\begin{bmatrix}0 & 1/\partial_y z_0 & 1\\ * & 0 & \partial_x z_0/\partial_y z_0\\ 1 & * & 2\partial_x z_0\end{bmatrix}[T_{0xz}]_{z0} \tag{17.26}$$

$$[\mathbf{S}_*]_{z0}=[\mathbf{S}_c]_{z0}=\mathbf{0} \tag{17.27}$$

17.4 (d) Independent *xx*-, *xy*-, *yz*-Components

Integral operators	∂_y^{-1}	∂_z^{-1}
Integration cones	K_y	K_z
Generative cone vectors	$\mathbf{e}_y$	$\mathbf{e}_z$
Model cone	K_{yz}	
A_0 – Normalization	$xx-yy-xy$	
Normalization directions	z	

$$\underbrace{\begin{bmatrix}\partial_z^2 & 0 & 0\\ 0 & -\partial_z^2 & 0\\ 0 & \partial_x\partial_z & -\partial_y\partial_z\end{bmatrix}}_{\mathcal{L}}\underbrace{\begin{bmatrix}A_{0yy}\\ A_{0xy}\\ A_{0xx}\end{bmatrix}}_{f}=\underbrace{\begin{bmatrix}T_{0xx}\\ T_{0xy}\\ T_{0yz}\end{bmatrix}}_{q} \tag{17.28}$$

$$\mathcal{L}^{-1}=\begin{bmatrix}1 & 0 & 0\\ 0 & -1 & 0\\ 0 & -\partial_x\partial_y^{-1} & -\partial_z\partial_y^{-1}\end{bmatrix}\cdot\partial_z^{-2} \tag{17.29}$$

$$\mathbf{A}_0 = \mathbf{A}_0^{\mathrm{T}} = \begin{bmatrix} 0 & 0 & 0 \\ 0 & \partial_z^{-2} & 0 \\ 0 & 0 & 0 \end{bmatrix} T_{0xx} + \begin{bmatrix} -\partial_x \partial_y^{-1} \partial_z^{-2} & -\partial_z^{-2} & 0 \\ * & 0 & 0 \\ 0 & 0 & 0 \end{bmatrix} T_{0xy} + \begin{bmatrix} -\partial_y^{-1} \partial_z^{-1} & 0 & 0 \\ 0 & 0 & 0 \\ 0 & 0 & 0 \end{bmatrix} T_{0yz} \tag{17.30}$$

$$\mathbf{B}_0 = \begin{bmatrix} 0 & -\partial_z^{-1} & 0 \\ 0 & 0 & 0 \\ 0 & \partial_x \partial_z^{-2} & 0 \end{bmatrix} T_{0xx} + \begin{bmatrix} \partial_z^{-1} & 0 & 0 \\ -\partial_x \partial_y^{-1} \partial_z^{-1} & -\partial_z^{-1} & 0 \\ 0 & \partial_y \partial_z^{-2} & 0 \end{bmatrix} T_{0xy} + \begin{bmatrix} 0 & 0 & 0 \\ -\partial_y^{-1} & 0 & 0 \\ \partial_z^{-1} & 0 & 0 \end{bmatrix} T_{0yz} \tag{17.31}$$

$$\mathbf{T}_0 = \mathbf{T}_0^{\mathrm{T}} = \begin{bmatrix} 1 & 0 & -\partial_x \partial_z^{-1} \\ 0 & 0 & 0 \\ * & 0 & \partial_x^2 \partial_z^{-2} \end{bmatrix} T_{0xx} + \begin{bmatrix} 0 & 1 & -\partial_y \partial_z^{-1} \\ 1 & -\partial_x \partial_y^{-1} & 0 \\ * & 0 & \partial_x \partial_y \partial_z^{-2} \end{bmatrix} T_{0xy} + \begin{bmatrix} 0 & 0 & 0 \\ 0 & -\partial_z \partial_y^{-1} & 1 \\ 0 & 1 & -\partial_y \partial_z^{-1} \end{bmatrix} T_{0yz} \tag{17.32}$$

$$\mathbf{S}_* = \mathbf{S}_d = \mathbf{S}_d^{\mathrm{T}} = \begin{bmatrix} 0 & 0 & -g_x \partial_z^{-1} \\ 0 & -g_y \partial_y^{-1} & 0 \\ * & 0 & (g_x \partial_x - g_z \partial_z) \cdot \partial_z^{-2} \end{bmatrix} \rho \tag{17.33}$$

$$[\mathbf{T}_0]_{z_0} = \left[\mathbf{T}_0^{\mathrm{T}}\right]_{z_0} = \begin{bmatrix} 1 & 0 & \partial_x z_0 \\ 0 & 0 & 0 \\ * & 0 & (\partial_x z_0)^2 \end{bmatrix} [T_{0xx}]_{z_0} + \begin{bmatrix} 0 & 1 & \partial_y z_0 \\ 1 & -\partial_x z_0 / \partial_y z_0 & 0 \\ * & 0 & \partial_x z_0 \cdot \partial_y z_0 \end{bmatrix} \left[T_{0xy}\right]_{z_0} + \begin{bmatrix} 0 & 0 & 0 \\ 0 & 1/\partial_y z_0 & 1 \\ 0 & 1 & \partial_y z_0 \end{bmatrix} \left[T_{0yz}\right]_{z_0} \tag{17.34}$$

$$[\mathbf{S}_*]_{z_0} = [\mathbf{S}_d]_{z_0} = \mathbf{0} \tag{17.35}$$

17.5 (e) Independent *xy*-, *yz*-, *xz*-Components

Integral operators	∂_x^{-1}	∂_y^{-1}	∂_z^{-1}
Integration cones	K_x	K_y	K_z
Generative cone vectors	$\mathbf{e}_x$	$\mathbf{e}_y$	$\mathbf{e}_z$
Model cone	K_{xyz}		
A_0 normalization	$xx-yy-xy$		
Normalization directions	z		

$$\underbrace{\begin{bmatrix} 0 & 0 & -\partial_z^2 \\ -\partial_y\partial_z & 0 & \partial_x\partial_z \\ 0 & -\partial_x\partial_z & \partial_y\partial_z \end{bmatrix}}_{\mathcal{L}} \underbrace{\begin{bmatrix} A_{0xx} \\ A_{0yy} \\ A_{0xy} \end{bmatrix}}_{f} = \underbrace{\begin{bmatrix} T_{0xy} \\ T_{0yz} \\ T_{0xz} \end{bmatrix}}_{q} \tag{17.36}$$

$$\mathcal{L}^{-1} = \begin{bmatrix} -\partial_x\partial_y^{-1} & -\partial_y^{-1}\partial_z & 0 \\ -\partial_y\partial_x^{-1} & 0 & -\partial_x^{-1}\partial_z \\ -1 & 0 & 0 \end{bmatrix} \cdot \partial_z^{-2} \tag{17.37}$$

$$\mathbf{A}_0 = \mathbf{A}_0^{\mathrm{T}} = -\begin{bmatrix} \partial_x\partial_y^{-1} & 1 & 0 \\ 1 & \partial_y\partial_x^{-1} & 0 \\ 0 & 0 & 0 \end{bmatrix} \cdot \partial_z^{-2} T_{0xy} - \begin{bmatrix} \partial_y^{-1}\partial_z^{-1} & 0 & 0 \\ 0 & 0 & 0 \\ 0 & 0 & 0 \end{bmatrix} T_{0yz} - \begin{bmatrix} 0 & 0 & 0 \\ 0 & \partial_x^{-1}\partial_z^{-1} & 0 \\ 0 & 0 & 0 \end{bmatrix} T_{0xz} \tag{17.38}$$

$$\mathbf{B}_0 = \begin{bmatrix} 1 & \partial_y\partial_x^{-1} & 0 \\ -\partial_x\partial_y^{-1} & -1 & 0 \\ 0 & 0 & 0 \end{bmatrix} \cdot \partial_z^{-1} T_{0xy} + \begin{bmatrix} 0 & 0 & 0 \\ -\partial_y^{-1} & 0 & 0 \\ \partial_z^{-1} & 0 & 0 \end{bmatrix} T_{0yz} + \begin{bmatrix} 0 & \partial_x^{-1} & 0 \\ 0 & 0 & 0 \\ 0 & -\partial_z^{-1} & 0 \end{bmatrix} T_{0xz} \tag{17.39}$$

$$\mathbf{T}_0 = \mathbf{T}_0^{\mathrm{T}} = \begin{bmatrix} -\partial_y\partial_x^{-1} & 1 & 0 \\ 1 & -\partial_x\partial_y^{-1} & 0 \\ 0 & 0 & 0 \end{bmatrix} T_{0xy} + \begin{bmatrix} 0 & 0 & 0 \\ 0 & -\partial_z\partial_y^{-1} & 1 \\ 0 & 1 & -\partial_y\partial_z^{-1} \end{bmatrix} T_{0yz} + \begin{bmatrix} -\partial_z\partial_x^{-1} & 0 & 1 \\ 0 & 0 & 0 \\ 1 & 0 & -\partial_x\partial_z^{-1} \end{bmatrix} T_{0xz} \tag{17.40}$$

$$\mathbf{S}_* = \mathbf{S}_e = \mathbf{S}_e^{\mathrm{T}} = \begin{bmatrix} -g_x\partial_x^{-1} & 0 & 0 \\ 0 & -g_y\partial_y^{-1} & 0 \\ 0 & 0 & -g_z\cdot\partial_z^{-1} \end{bmatrix}\rho \tag{17.41}$$

$$\begin{aligned}[\mathbf{T}_0]_{z0} = \left[\mathbf{T}_0^{\mathrm{T}}\right]_{z0} = &\begin{bmatrix} -\partial_{y}z_0/\partial_{x}z_0 & 1 & 0 \\ 1 & -\partial_{x}z_0/\partial_{y}z_0 & 0 \\ 0 & 0 & 0 \end{bmatrix}\left[T_{0xy}\right]_{z0} \\ &+ \begin{bmatrix} 0 & 0 & 0 \\ 0 & 1/\partial_{y}z_0 & 1 \\ 0 & 1 & \partial_{y}z_0 \end{bmatrix}\left[T_{0yz}\right]_{z0} + \begin{bmatrix} 1/\partial_{x}z_0 & 0 & 1 \\ 0 & 0 & 0 \\ 1 & 0 & \partial_{x}z_0 \end{bmatrix}\left[T_{0xz}\right]_{z0}\end{aligned} \tag{17.42}$$

$$[\mathbf{S}_*]_{z0} = [\mathbf{S}_e]_{z0} = \mathbf{0} \tag{17.43}$$

17.6 (f) Independent Deviatoric *xx*-, *yy*-, *xy*-Components

Integral operators	∂_z^{-1}	$\square_z^{-1}$
Integration cones	K_z	$\mathrm{K}_z^{\odot}$
Generative cone vectors	$\mathbf{e}_z$	$\mathbf{e}_x\cdot\cos\varphi+\mathbf{e}_y\cdot\sin\varphi+\mathbf{e}_z$ $0\le\varphi<2\pi$
Model cone	$\mathrm{K}_z^{\odot}$	
A_0 Normalization	$xx-yy-xy$	
Normalization directions	z	

$$\underbrace{\begin{bmatrix} -(\partial_y^2+\partial_z^2)/3 & (-\partial_x^2+2\partial_z^2)/3 & 2\partial_x\partial_y/3 \\ (-\partial_y^2+2\partial_z^2)/3 & -(\partial_x^2+\partial_z^2)/3 & 2\partial_x\partial_y/3 \\ 0 & 0 & -\partial_z^2 \end{bmatrix}}_{\mathcal{L}}\underbrace{\begin{bmatrix} A_{0xx} \\ A_{0yy} \\ A_{0xy} \end{bmatrix}}_{f} = \underbrace{\begin{bmatrix} T'_{0xx} \\ T'_{0yy} \\ T_{0xy} \end{bmatrix}}_{q} \tag{17.44}$$

$$\mathcal{L}^{-1} = \begin{bmatrix} \partial_x^2+\partial_z^2 & -\partial_x^2+2\partial_z^2 & 2\partial_x\partial_y \\ -\partial_y^2+2\partial_z^2 & \partial_y^2+\partial_z^2 & 2\partial_x\partial_y \\ 0 & 0 & -\square_z \end{bmatrix}\cdot\partial_z^{-2}\square_z^{-1} \tag{17.45}$$

$$\mathbf{A}_0 = \mathbf{A}_0^{\mathrm{T}} = \begin{bmatrix} \partial_x^2+\partial_z^2 & 0 & 0 \\ 0 & -\partial_y^2+2\partial_z^2 & 0 \\ 0 & 0 & 0 \end{bmatrix} \partial_z^{-2}\square_z^{-1} T'_{0xx}$$

$$+ \begin{bmatrix} -\partial_x^2+2\partial_z^2 & 0 & 0 \\ 0 & \partial_y^2+\partial_z^2 & 0 \\ 0 & 0 & 0 \end{bmatrix} \partial_z^{-2}\square_z^{-1} T'_{0yy} + \begin{bmatrix} 2\partial_x\partial_y & -\square_z & 0 \\ * & 2\partial_x\partial_y & 0 \\ 0 & 0 & 0 \end{bmatrix} \partial_z^{-2}\square_z^{-1} T_{0xy} \tag{17.46}$$

$$\mathbf{B}_0 = \begin{bmatrix} 0 & \left(\partial_y^2-2\partial_z^2\right) & 0 \\ \left(\partial_x^2+\partial_z^2\right) & 0 & 0 \\ -\partial_y\left(\partial_x^2+\partial_z^2\right)\partial_z^{-1} & -\partial_x\left(\partial_y^2-2\partial_z^2\right)\partial_z^{-1} & 0 \end{bmatrix} \partial_z^{-1}\square_z^{-1} T'_{0xx}$$

$$+ \begin{bmatrix} 0 & -\left(\partial_y^2+\partial_z^2\right) & 0 \\ -\left(\partial_x^2-2\partial_z^2\right) & 0 & 0 \\ \partial_y\left(\partial_x^2-2\partial_z^2\right)\partial_z^{-1} & \partial_x\left(\partial_y^2+\partial_z^2\right)\partial_z^{-1} & 0 \end{bmatrix} \partial_z^{-1}\square_z^{-1} T'_{0yy}$$

$$+ \begin{bmatrix} \square_z & -2\partial_x\partial_y & 0 \\ 2\partial_x\partial_y & -\square_z & 0 \\ \partial_x\square_x\partial_z^{-1} & -\partial_y\square_y\partial_z^{-1} & 0 \end{bmatrix} \partial_z^{-1}\square_z^{-1} T_{0xy} \tag{17.47}$$

$$\mathbf{T}_0 = \mathbf{T}_0^{\mathrm{T}} = \begin{bmatrix} 2\partial_z^2-\partial_y^2 & 0 & \partial_x\partial_z^{-1}\left(\partial_y^2-2\partial_z^2\right) \\ 0 & \partial_x^2+\partial_z^2 & -\partial_y\partial_z^{-1}\left(\partial_x^2+\partial_z^2\right) \\ * & * & 2\partial_x^2+\partial_y^2 \end{bmatrix} \square_z^{-1} T'_{0xx}$$

$$+ \begin{bmatrix} \partial_y^2+\partial_z^2 & 0 & -\partial_x\partial_z^{-1}\left(\partial_y^2+\partial_z^2\right) \\ 0 & 2\partial_z^2-\partial_x^2 & \partial_y\partial_z^{-1}\left(\partial_x^2-2\partial_z^2\right) \\ * & * & \partial_x^2+2\partial_y^2 \end{bmatrix} \square_z^{-1} T'_{0yy}$$

$$+ \begin{bmatrix} 2\partial_x\partial_y & \square_z & \partial_y\partial_z^{-1}\square_y \\ * & 2\partial_x\partial_y & \partial_x\partial_z^{-1}\square_x \\ * & * & 2\partial_x\partial_y \end{bmatrix} \square_z^{-1} T_{0xy} \tag{17.48}$$

$$\mathbf{S}_* = \mathbf{S}_f = \mathbf{S}_f^{\mathrm{T}} = \begin{bmatrix} g_x\partial_x+g_y\partial_y-g_z\partial_z & 0 & \left(g_x\partial_y-g_y\partial_x\right)\partial_y\partial_z^{-1}-g_x\partial_z+g_z\partial_x \\ * & g_x\partial_x+g_y\partial_y-g_z\partial_z & \left(g_y\partial_x-g_x\partial_y\right)\partial_x\partial_z^{-1}-g_y\partial_z+g_z\partial_y \\ * & * & g_x\partial_x+g_y\partial_y-g_z\partial_z \end{bmatrix} \square_z^{-1}\rho \tag{17.49}$$

$$[\mathbf{T}_0]_{z_0} = \left[\mathbf{T}_0^{\mathrm{T}}\right]_{z_0} = \begin{bmatrix} 2-\left(\partial_y z_0\right)^2 & 0 & \partial_x z_0\cdot\left[2-\left(\partial_y z_0\right)^2\right] \\ 0 & 1+\left(\partial_x z_0\right)^2 & \partial_y z_0\cdot\left[1+\left(\partial_x z_0\right)^2\right] \\ * & * & 2\left(\partial_x z_0\right)^2+\left(\partial_y z_0\right)^2 \end{bmatrix} \cdot \frac{\left[T'_{0xx}\right]_{z_0}}{N}$$

$$+ \begin{bmatrix} 1+\left(\partial_y z_0\right)^2 & 0 & \partial_x z_0\cdot\left[1+\left(\partial_y z_0\right)^2\right] \\ 0 & 2-\left(\partial_x z_0\right)^2 & \partial_y z_0\cdot\left[2-\left(\partial_x z_0\right)^2\right] \\ * & * & \left(\partial_x z_0\right)^2+2\left(\partial_y z_0\right)^2 \end{bmatrix} \cdot \frac{\left[T'_{0yy}\right]_{z_0}}{N}$$

$$+ \begin{bmatrix} 2\cdot\partial_x z_0\cdot\partial_y z_0 & N & \partial_y z_0\cdot(1+M) \\ * & 2\cdot\partial_x z_0\cdot\partial_y z_0 & \partial_x z_0\cdot(1-M) \\ * & * & 2\cdot\partial_x z_0\cdot\partial_y z_0 \end{bmatrix} \cdot \frac{\left[T_{0xy}\right]_{z_0}}{N} \tag{17.50}$$

$$M \overset{\text{def.}}{=} \left(\partial_x z_0\right)^2 - \left(\partial_y z_0\right)^2 \tag{17.51}$$

$$N \overset{\text{def.}}{=} \left[1 - (\partial_x z_0)^2 - (\partial_y z_0)^2\right] \tag{17.52}$$

$$[\mathbf{S}_*]_{z_0} = [\mathbf{S}_f]_{z_0} = \mathbf{0} \tag{17.53}$$

17.7 (g) Independent Deviatoric *xx*-, *yy*-, *xz*-Components

Integral operators	∂_y^{-1}	$\square_y^{-1}$
Integration cones	K_y	$\mathrm{K}_y^{\odot}$
Generative cone vectors	$\mathbf{e}_y$	$\mathbf{e}_x \cdot \cos\varphi + \mathbf{e}_z \cdot \sin\varphi + \mathbf{e}_y$ $0 \leq \varphi < 2\pi$
Model cone	$\mathrm{K}_y^{\odot}$	
A_0 normalization	$xx - zz - xz$	
Normalization directions	y	

$$\underbrace{\frac{1}{3}\begin{bmatrix} -\partial_y^2 - \partial_z^2 & -\partial_x^2 + 2\partial_y^2 & 2\partial_x\partial_z \\ -\partial_y^2 + 2\partial_z^2 & 2\partial_x^2 - \partial_y^2 & -4\partial_x\partial_z \\ 0 & 0 & -3\partial_y^2 \end{bmatrix}}_{\mathcal{L}} \underbrace{\begin{bmatrix} A_{0xx} \\ A_{0zz} \\ A_{0xz} \end{bmatrix}}_{\boldsymbol{f}} = \underbrace{\begin{bmatrix} T'_{0xx} \\ T'_{0yy} \\ T_{0xz} \end{bmatrix}}_{\boldsymbol{q}} \tag{17.54}$$

$$\mathcal{L}^{-1} = \begin{bmatrix} 2\partial_x^2 - \partial_y^2 & \partial_x^2 - 2\partial_y^2 & 2\partial_x\partial_z \\ \partial_y^2 - 2\partial_z^2 & -\partial_y^2 - \partial_z^2 & 2\partial_x\partial_z \\ 0 & 0 & -\square_y \end{bmatrix} \cdot \partial_y^{-2}\square_y^{-1} \tag{17.55}$$

$$\begin{aligned} \mathbf{A}_0 = \mathbf{A}_0^{\mathrm{T}} = & \begin{bmatrix} 2\partial_x^2 - \partial_y^2 & 0 & 0 \\ 0 & 0 & 0 \\ 0 & 0 & \partial_y^2 - 2\partial_z^2 \end{bmatrix} \partial_y^{-2}\square_y^{-1} T'_{0xx} \\ & + \begin{bmatrix} \partial_x^2 - 2\partial_y^2 & 0 & 0 \\ 0 & 0 & 0 \\ 0 & 0 & -\partial_y^2 - \partial_z^2 \end{bmatrix} \partial_y^{-2}\square_y^{-1} T'_{0yy} + \begin{bmatrix} 2\partial_x\partial_z & 0 & -\square_y \\ 0 & 0 & 0 \\ * & 0 & 2\partial_x\partial_z \end{bmatrix} \partial_y^{-2}\square_y^{-1} T_{0xz} \end{aligned} \tag{17.56}$$

$$\mathbf{B}_0=\begin{bmatrix} 0 & 0 & \left(\partial_y^2-2\partial_z^2\right) \\ \partial_z\left(2\partial_x^2-\partial_y^2\right)\partial_y^{-1} & 0 & -\partial_x\left(\partial_y^2-2\partial_z^2\right)\partial_y^{-1} \\ \left(-2\partial_x^2+\partial_y^2\right) & 0 & 0 \end{bmatrix}\partial_y^{-1}\Box_y^{-1}T'_{0xx}$$
$$+\begin{bmatrix} 0 & 0 & -\left(\partial_y^2+\partial_z^2\right) \\ \partial_z\left(\partial_x^2-2\partial_y^2\right)\partial_y^{-1} & 0 & \partial_x\left(\partial_y^2+\partial_z^2\right)\partial_y^{-1} \\ \left(-\partial_x^2+2\partial_y^2\right) & 0 & 0 \end{bmatrix}\partial_y^{-1}\Box_y^{-1}T'_{0yy}$$
$$+\begin{bmatrix} -\Box_y & 0 & 2\partial_x\partial_z \\ -\partial_x\Box_x\partial_y^{-1} & 0 & \partial_z\Box_z\partial_y^{-1} \\ -2\partial_x\partial_z & 0 & \Box_y \end{bmatrix}\partial_y^{-1}\Box_y^{-1}T_{0xz} \tag{17.57}$$

$$\mathbf{T}_0=\mathbf{T}_0^{\mathrm{T}}=\begin{bmatrix} \partial_y^2-2\partial_z^2 & \partial_x\left(2\partial_z^2-\partial_y^2\right)\partial_y^{-1} & 0 \\ * & \partial_x^2-\partial_z^2 & \partial_z\left(\partial_y^2-2\partial_x^2\right)\partial_y^{-1} \\ 0 & * & 2\partial_x^2-\partial_y^2 \end{bmatrix}\Box_y^{-1}T'_{0xx}$$
$$+\begin{bmatrix} -\partial_y^2-\partial_z^2 & +\partial_x\left(\partial_y^2+\partial_z^2\right)\partial_y^{-1} & 0 \\ * & -\partial_x^2-2\partial_z^2 & \partial_z\left(2\partial_y^2-\partial_x^2\right)\partial_y^{-1} \\ * & * & \partial_x^2-2\partial_y^2 \end{bmatrix}\Box_y^{-1}T'_{0yy}$$
$$+\begin{bmatrix} 2\partial_x\partial_z & \partial_z\Box_z\partial_y^{-1} & \Box_y \\ * & 2\partial_x\partial_z & \partial_x\Box_x\partial_y^{-1} \\ * & * & 2\partial_x\partial_z \end{bmatrix}\Box_y^{-1}T_{0xz} \tag{17.58}$$

$$\mathbf{S}_*=\mathbf{S}_g=\mathbf{S}_g^{\mathrm{T}}=\begin{bmatrix} g_x\partial_x-g_y\partial_y+g_z\partial_z & \left(g_x\partial_z-g_z\partial_x\right)\cdot\partial_z\partial_y^{-1}-g_x\partial_y+g_y\partial_x & 0 \\ * & g_x\partial_x-g_y\partial_y+g_z\partial_z & \left(g_z\partial_x-g_x\partial_z\right)\cdot\partial_x\partial_y^{-1}-g_z\partial_y+g_y\partial_z \\ 0 & * & g_x\partial_x-g_y\partial_y+g_z\partial_z \end{bmatrix}\cdot\Box_y^{-1}\rho \tag{17.59}$$

$$[\mathbf{T}_0]_{y_0}=\left[\mathbf{T}_0^{\mathrm{T}}\right]_{y_0}=\begin{bmatrix} 1-2(\partial_z y_0)^2 & \partial_x y_0\left[1-2(\partial_z y_0)^2\right] & 0 \\ * & M & \partial_z y_0\cdot\left[2(\partial_x y_0)^2-1\right] \\ 0 & * & 2(\partial_x y_0)^2-1 \end{bmatrix}\cdot\frac{\left[T'_{0xx}\right]_{y_0}}{N}$$
$$-\begin{bmatrix} 1+(\partial_z y_0)^2 & \partial_x y_0\left[1+(\partial_z y_0)^2\right] & 0 \\ * & (\partial_x y_0)^2+2(\partial_z y_0)^2 & \partial_z y_0\left[2-(\partial_x y_0)^2\right] \\ 0 & * & 2-(\partial_x y_0)^2 \end{bmatrix}\cdot\frac{\left[T'_{0yy}\right]_{y_0}}{N}$$
$$+\begin{bmatrix} 2\cdot\partial_x y_0\cdot\partial_z y_0 & \partial_z y_0\cdot(1+M) & N \\ * & 2\cdot\partial_x y_0\cdot\partial_z y_0 & \partial_x y_0\cdot(1-M) \\ * & * & 2\cdot\partial_x y_0\cdot\partial_z y_0 \end{bmatrix}\cdot\frac{\left[T_{0xz}\right]_{y_0}}{N} \tag{17.60}$$

$$M\overset{\text{def.}}{=}(\partial_x y_0)^2-(\partial_z y_0)^2 \tag{17.61}$$

$$N\overset{\text{def.}}{=}\left[1-(\partial_x y_0)^2-(\partial_z y_0)^2\right] \tag{17.62}$$

$$[\mathbf{S}_*]_{y_0}=\left[\mathbf{S}_g\right]_{y_0}=\mathbf{0} \tag{17.63}$$

17.8 (h) Independent Deviatoric *xx, xy, yz* Components

Integral operators	∂_y^{-1}	$\left(\partial_x+\sqrt{2}\partial_z\right)^{-1}$	$\left(-\partial_x+\sqrt{2}\partial_z\right)^{-1}$
Integration cones	K_y	(unmarked)	(unmarked)
Generative cone vectors	$\mathbf{e}_y$	$\mathbf{e}_x+\sqrt{2}\mathbf{e}_z$	$-\mathbf{e}_x+\sqrt{2}\mathbf{e}_z$
Model cone		K'_{yxz}	

or

Integral operators	∂_y^{-1}	$\left(\partial_x+\sqrt{2}\partial_z\right)^{-1}$	$\left(\partial_x-\sqrt{2}\partial_z\right)^{-1}$
Integration cones	K_y	(unmarked)	(unmarked)
Generative cone vectors	$\mathbf{e}_y$	$\mathbf{e}_x+\sqrt{2}\mathbf{e}_z$	$\mathbf{e}_x-\sqrt{2}\mathbf{e}_z$
Model cone		K''_{yxz}	

A_0 normalization	$xx-zz-xz$
Normalization directions	y

$$\underbrace{\begin{bmatrix} -\partial_z^2-\partial_y^2 & 2\partial_y^2-\partial_x^2 & 2\partial_x\partial_z \\ 0 & -\partial_x\partial_y & \partial_y\partial_z \\ -\partial_y\partial_z & 0 & \partial_x\partial_y \end{bmatrix}}_{\mathcal{L}}\underbrace{\begin{bmatrix} A_{0xx} \\ A_{0zz} \\ A_{0xz} \end{bmatrix}}_{\mathbf{f}}=\underbrace{\begin{bmatrix} 3T'_{0xx} \\ T_{0xy} \\ T_{0yz} \end{bmatrix}}_{\mathbf{q}} \tag{17.64}$$

$$\partial \overset{\text{def.}}{=} \partial_x^2-2\partial_z^2 \tag{17.65}$$

$$\begin{aligned}\partial^{-1}&=-\left(\partial_x+\sqrt{2}\partial_z\right)^{-1}\left(-\partial_x+\sqrt{2}\partial_z\right)^{-1}\\&=\left(\partial_x+\sqrt{2}\partial_z\right)^{-1}\left(\partial_x-\sqrt{2}\partial_z\right)^{-1}\end{aligned} \tag{17.66}$$

$$\mathcal{L}^{-1}=\begin{bmatrix} -\partial_x^2\partial_y & -\partial_x\left(2\partial_y^2-\partial_x^2\right) & \partial_z\left(2\partial_y^2+\partial_x^2\right) \\ -\partial_z^2\partial_y & \partial_x\left(\partial_z^2-\partial_y^2\right) & \partial_z\left(\partial_z^2+\partial_y^2\right) \\ -\partial_x\partial_y\partial_z & -\partial_z\left(2\partial_y^2-\partial_x^2\right) & \partial_x\left(\partial_z^2+\partial_y^2\right) \end{bmatrix}\partial_y^{-3}\partial^{-1} \tag{17.67}$$

$$
\begin{aligned}
\mathbf{A}_0=\mathbf{A}_0^{\mathrm{T}}=&\begin{bmatrix} -\partial_x^2 & 0 & -\partial_x\partial_z \\ 0 & 0 & 0 \\ * & 0 & -\partial_z^2 \end{bmatrix} 3\partial_y^{-2}\partial^{-1}T'_{0xx} \\
&+\begin{bmatrix} -\partial_x\left(2\partial_y^2-\partial_x^2\right) & 0 & -\partial_z\left(2\partial_y^2-\partial_x^2\right) \\ 0 & 0 & 0 \\ * & 0 & \partial_x\left(\partial_z^2-\partial_y^2\right) \end{bmatrix} \partial_y^{-3}\partial^{-1}T_{0xy} \\
&+\begin{bmatrix} \partial_z\left(2\partial_y^2+\partial_x^2\right) & 0 & \partial_x\left(\partial_z^2+\partial_y^2\right) \\ 0 & 0 & 0 \\ * & 0 & \partial_z\left(\partial_z^2+\partial_y^2\right) \end{bmatrix} \partial_y^{-3}\partial^{-1}T_{0yz}
\end{aligned}
\tag{17.68}
$$

$$
\begin{aligned}
\mathbf{B}_0=&\begin{bmatrix} -\partial_x\partial_z & 0 & -\partial_z^2 \\ 0 & 0 & 0 \\ \partial_x^2 & 0 & \partial_x\partial_z \end{bmatrix} 3\partial_y^{-1}\partial^{-1}T'_{0xx} \\
&+\begin{bmatrix} -\partial_z\left(2\partial_y^2-\partial_x^2\right) & 0 & \partial_x\left(\partial_z^2-\partial_y^2\right) \\ 0 & 0 & \partial_y\left(\partial_x^2-2\partial_z^2\right) \\ \partial_x\left(2\partial_y^2-\partial_x^2\right) & 0 & \partial_z\left(2\partial_y^2-\partial_x^2\right) \end{bmatrix} \partial_y^{-2}\partial^{-1}T_{0xy} \\
&+\begin{bmatrix} \partial_x\left(\partial_y^2+\partial_z^2\right) & 0 & \partial_z\left(\partial_y^2+\partial_z^2\right) \\ -\partial_y\left(\partial_x^2-2\partial_z^2\right) & 0 & 0 \\ -\partial_z\left(\partial_x^2+2\partial_y^2\right) & 0 & -\partial_x\left(\partial_y^2+\partial_z^2\right) \end{bmatrix} \partial_y^{-2}\partial^{-1}T_{0yz}
\end{aligned}
\tag{17.69}
$$

$$
\begin{aligned}
\mathbf{T}_0=\mathbf{T}_0^{\mathrm{T}}=3&\begin{bmatrix} -\partial_z^2 & 0 & \partial_x\partial_z \\ 0 & 0 & 0 \\ * & 0 & -\partial_x^2 \end{bmatrix} \partial^{-1}T'_{0xx} \\
&+\begin{bmatrix} \partial_x\left(\partial_z^2-\partial_y^2\right)\partial_y^{-1} & \partial & \partial_z\left(2\partial_y^2-\partial_x^2\right)\partial_y^{-1} \\ * & -\partial_x\partial\partial_y^{-1} & 0 \\ * & 0 & \partial_x\left(\partial_x^2-2\partial_y^2\right)\partial_y^{-1} \end{bmatrix} \partial^{-1}T_{0xy} \\
&+\begin{bmatrix} \partial_z\left(\partial_y^2+\partial_z^2\right)\partial_y^{-1} & 0 & -\partial_x\left(\partial_y^2+\partial_z^2\right)\partial_y^{-1} \\ 0 & -\partial_z\partial\partial_y^{-1} & \partial \\ * & * & \partial_z\left(\partial_x^2+2\partial_y^2\right)\partial_y^{-1} \end{bmatrix} \partial^{-1}T_{0yz}
\end{aligned}
\tag{17.70}
$$

$$
\mathbf{S}_*=\mathbf{S}_h=\mathbf{S}_h^{\mathrm{T}}=\begin{bmatrix} \partial_z\partial_y^{-1}\left(\mathrm{g}_y\partial_z-\mathrm{g}_z\partial_y\right)-\mathrm{g}_x\partial_x+2\mathrm{g}_z\partial_z & 0 & \partial_x\partial_y^{-1}\left(\mathrm{g}_z\partial_y-\mathrm{g}_y\partial_z\right)+2\left(\mathrm{g}_x\partial_z-\mathrm{g}_z\partial_x\right) \\ 0 & -\mathrm{g}_y\partial_y^{-1}\partial & 0 \\ * & 0 & \partial_x\partial_y^{-1}\left(\mathrm{g}_y\partial_x-\mathrm{g}_x\partial_y\right)-\mathrm{g}_x\partial_x+2\mathrm{g}_z\partial_z \end{bmatrix} \partial^{-1}\rho
\tag{17.71}
$$

$$
\begin{aligned}
[\mathbf{T}_0]_{y_0}=\left[\mathbf{T}_0^{\mathrm{T}}\right]_{y_0}=&\begin{bmatrix} -3(\partial_z y_0)^2 & 0 & 3\partial_x y_0\cdot\partial_z y_0 \\ 0 & 0 & 0 \\ * & 0 & -3(\partial_x y_0)^2 \end{bmatrix} \cdot \frac{\left[T'_{0xx}\right]_{y_0}}{N} \\
&+\begin{bmatrix} \partial_x y_0\left[1-(\partial_z y_0)^2\right] & N & \partial_z y_0\left[-2+(\partial_x y_0)^2\right] \\ * & N\partial_x y_0 & 0 \\ * & 0 & \partial_x y_0\left[2-(\partial_x y_0)^2\right] \end{bmatrix} \cdot \frac{\left[T_{0xy}\right]_{y_0}}{N} \\
&+\begin{bmatrix} \partial_z y_0\left[-1-(\partial_z y_0)^2\right] & 0 & \partial_x y_0\left[1+(\partial_z y_0)^2\right] \\ 0 & N\partial_z y_0 & N \\ * & * & \partial_z y_0\left[-2-(\partial_x y_0)^2\right] \end{bmatrix} \cdot \frac{\left[T_{0yz}\right]_{y_0}}{N}
\end{aligned}
\tag{17.72}
$$

$$N \overset{\text{def.}}{=} (\partial_x y_0)^2 - 2(\partial_z y_0)^2 \tag{17.73}$$

$$[\mathbf{S}_*]_{y_0} = [\mathbf{S}_h]_{y_0} = \mathbf{0} \tag{17.74}$$

18 Transformations

Abstract

The following transformations concern the general solution of the balance and boundary conditions in models with three independent stress components. These transformations serve to represent the general solution by integrals. Thereby, by differentiating the step function, intermediate results occur, which contain the delta function.

The following transformations concern the general solution **S** of the balance and boundary conditions in models with three independent stress components.[1] These transformations serve to represent the general solution by integrals.

This general solution is obtained by differentiating and integrating the three independent stress components and the ice density by applying differential and integral operators to these functions,[2] all operators being interchangeable. The integral representation of the general solution is obtained by first applying the differential operators and then the integral operators.

All relations in this chapter are valid if the free ice surface Σ can be defined both in the form $z = z_0(x, y)$ and $y = y_0(x, z)$and $x = x_0(y, z)$ as well. If one of these conditions is not fulfilled, the corresponding relations are omitted. If, for example, there is no representation in the form $x = x_0(y, z)$, then all relations in which x_0 occurs have to be omitted in the following.

---

[1] See Sect. 8.2.

[2] S. Chap. 17.

P. Halfar, *Stresses in glaciers*, https://doi.org/10.1007/978-3-662-66024-9_18

18.1 Spatial Domain of Definition

In the following, the spatial domain of definition Ω_{def} in which the calculations take place is described once again.[3]

The free ice surface Σ must be transverse and synchronous to the model cone, i.e. all cone rays of the model cone emanating from Σ must run into the air.[4] The free ice surface Σ can be defined by a function $z_0(x, y)$, or by a function $y_0(x, z)$ or by a function $x_0(y, z)$ (18.1). If the surface Σ can be defined simultaneously by several such functions, then relations (18.2) exist between these functions, and relations between the derivatives of these functions (18.3)–(18.6) exist on the surface Σ.

The considered glacier area Ω extends from the surface Σ in negative z- resp. x- resp. y-direction, if Σ can be defined by a function $z_0(x, y)$ resp. . $x_0(y, z)$ resp. $y_0(x, z)$. (18.7) Ω must be compatible with Σ and the model cone, i.e. all rays of the model cone originating from Ω must run uninterrupted in Ω until they hit Σ.[5] The domain of definition Ω_{def} (18.9) is larger than the glacier domain Ω and additionally contains the external domain Ω_{ext}, which lies beyond the surface Σ in positive z- or x- or y-direction (18.8). This external domain is generated by all model cones with the tip on the surface Σ and was introduced only to define the integral operators whose integration cones run into this external domain. On this external domain all functions and distributions disappear.

A formal problem may be that the functions $z_0(x, y)$ resp. $y_0(x, z)$ resp. $x_0(y, z)$, which describe the free surface Σ, are defined everywhere in the considered glacier domain Ω, but not everywhere in the external domain Ω_{ext}, so that the corresponding relations (18.8) are not initially explained everywhere in the external domain. This formal problem is solved by defining these functions in the external domain also where they are not explained so far, so that the corresponding relations (18.8) are valid in the whole external domain Ω_{ext} and thus also in the whole domain of definition Ω_{def}.[6]

$$z = z_0(x, y); \quad y = y_0(x, z); \quad x = x_0(y, z) \tag{18.1}$$

$$z \overset{\text{id.}}{=} z_0[x, y_0(x, z)] \overset{\text{id.}}{=} z_0[x_0(y, z), y] \tag{18.2}$$

[3] See Sect. 9.1.2, No. 1, 6.

[4] See Sect. 3.4.1, No. 1.

[5] See Sect. 3.4.1, point 2.

[6] So the area described by these functions z_0 resp. y_0 resp. x_0 is extended in such a way that the extended area lies in z- resp. x- resp. y-direction below all points of Ω_{ext}, so that the relations (18.8) are valid everywhere in Ω_{ext}.

$$[\partial_x y_0]_\Sigma = -\left[\frac{\partial_x z_0}{\partial_y z_0}\right]_\Sigma \tag{18.3}$$

$$[\partial_z y_0]_\Sigma = \frac{1}{[\partial_y z_0]} \tag{18.4}$$

$$[\partial_y x_0]_\Sigma = -\left[\frac{\partial_y z_0}{\partial_x z_0}\right]_\Sigma \tag{18.5}$$

$$[\partial_z x_0]_\Sigma = \frac{1}{[\partial_x z_0]_\Sigma} \tag{18.6}$$

$$z \le z_0(x,y); \quad y \le y_0(x,z); \quad x \le x_0(y,z); \quad \mathbf{r} \in \Omega \tag{18.7}$$

$$z_0(x,y) < z; \quad y_0(x,z) < y; \quad x_0(y,z) < x; \quad \mathbf{r} \in \Omega_{\text{ext}} \tag{18.8}$$

$$\Omega_{\text{def}} = \Omega \cup \Omega_{\text{ext}} \tag{18.9}$$

18.2 Heaviside and Delta Function

Some calculation rules follow for the Heaviside or step function θ, for the delta function δ with its derivatives δ' and δ'', and for the products of a smooth function ψ and derivatives of the delta function. The step function θ has a value of 1 in the glacier region Ω and vanishes in the external region Ω_{ext} beyond the free surface. The delta function and its derivatives arise from derivatives of this step function. These functions and distributions are used to build other functions and distributions, which vanish in the external domain Ω_{ext}.

We assume that the free surface can be defined by a function $z_0(x, y)$ in any case, and in some cases still by a function $y_0(x, z)$ or a function $x_0(y, z)$.

—

$$\theta \overset{\text{def}}{=} \theta(z_0 - z) = \theta(y_0 - y) = \theta(x_0 - x) = \begin{cases} 1; \mathbf{r}\in\Omega \\ 0; \mathbf{r}\in\Omega_{\text{ext}} \end{cases} \tag{18.10}$$

$$\delta \overset{\text{def.}}{=} \delta(z_0 - z) = \theta'(z_0 - z) = -\partial_z \theta(z_0 - z)$$

$$= -\frac{\delta(y_0 - y)}{\partial_y z_0} = -\frac{\delta(x_0 - x)}{\partial_x z_0} \tag{18.11}$$

$$\delta(y_0 - y) = \theta'(y_0 - y) = -\partial_y\theta = -\delta(z_0 - z) \cdot \partial_y z_0 \tag{18.12}$$

$$\delta(x_0 - x) = -\delta(z_0 - z) \cdot \partial_x z_0 \tag{18.13}$$

$$\delta' \overset{\text{def.}}{=} \delta'(z_0 - z) = -\partial_z \delta(z_0 - z) \tag{18.14}$$

$$\delta'' \overset{\text{def.}}{=} \delta''(z_0 - z) = \partial_z^2 \delta(z_0 - z) \tag{18.15}$$

$$\begin{aligned}\delta' \cdot \psi &= -\partial_z(\delta \cdot \psi) + \delta \cdot \partial_z \psi \\ &= \partial_y\left(\frac{\delta \cdot \psi}{\partial_y z_0}\right) - \delta \cdot \partial_y\left(\frac{\psi}{\partial_y z_0}\right)\end{aligned} \tag{18.16}$$

$$\delta'' \cdot \psi = \partial_z^2(\delta \cdot \psi) - 2\partial_z(\delta \cdot \partial_z \psi) + \delta \cdot \partial_z^2 \psi \tag{18.17}$$

18.3 Derivatives

The derivatives of the product of a smooth function ψ and the step function θ are transformed. This also gives products of derivatives of the delta function and of smooth functions, which are replaced by expressions, (18.16) and (18.17) in which only total z-derivatives of products with the delta function occur. The transformed expressions consist of a product of the step function and a smooth function, of a product of the delta function and a smooth function,[7] and of first and second z-derivatives of such products (18.18). The function symbols θ and δ occurring without arguments in this process have already been defined (18.10) and (18.11).

$$\theta \cdot \psi_1 + \delta \cdot \psi_2 + \partial_z(\delta \cdot \psi_3) + \partial_z^2(\delta \cdot \psi_4) \tag{18.18}$$

* * *

$$\partial_y(\theta \cdot \psi) = \theta \cdot \partial_y \psi + \delta \cdot \partial_y z_0 \cdot \psi \tag{18.19}$$

[7] Here it is assumed that the function is also $z_0(x, y)$ smooth.

$$\partial_x(\theta \cdot \psi) = \theta \cdot \partial_x\psi + \delta \cdot \partial_x z_0 \cdot \psi \tag{18.20}$$

$$\partial_z(\theta \cdot \psi) = \theta \cdot \partial_z\psi - \delta \cdot \psi \tag{18.21}$$

* * *

$$\begin{aligned}\partial_x\partial_y(\theta \cdot \psi) = \theta \cdot \partial_x\partial_y\psi + \delta \cdot [\partial_z\psi \cdot \partial_x z_0 \cdot \partial_y z_0 \\ +\partial_x\psi \cdot \partial_y z_0 + \partial_y\psi \cdot \partial_x z_0 + \psi \cdot \partial_x\partial_y z_0] - \partial_z[\delta \cdot \psi \cdot \partial_x z_0 \cdot \partial_y z_0]\end{aligned} \tag{18.22}$$

$$\begin{aligned}\partial_y^2(\theta \cdot \psi) = \theta \cdot \partial_y^2\psi + \delta \cdot \left[\partial_z\psi \cdot (\partial_y z_0)^2\right. \\ \left.+2 \cdot \partial_y\psi \cdot \partial_y z_0 + \psi \cdot \partial_y^2 z_0\right] - \partial_z\left[\delta \cdot \psi \cdot (\partial_y z_0)^2\right]\end{aligned} \tag{18.23}$$

$$\begin{aligned}\partial_x^2(\theta \cdot \psi) = \theta \cdot \partial_x^2\psi + \delta \cdot \left[\partial_z\psi \cdot (\partial_x z_0)^2\right. \\ \left.+2 \cdot \partial_x\psi \cdot \partial_x z_0 + \psi \cdot \partial_x^2 z_0\right] - \partial_z\left[\delta \cdot \psi \cdot (\partial_x z_0)^2\right]\end{aligned} \tag{18.24}$$

$$\partial_y\partial_z(\theta \cdot \psi) = \theta \cdot \partial_z\partial_y\psi - \delta \cdot \partial_y\psi + \partial_z(\delta \cdot \psi \cdot \partial_y z_0) \tag{18.25}$$

$$\partial_z^2(\theta \cdot \psi) = \theta \cdot \partial_z^2\psi - 2\delta \cdot \partial_z\psi + \delta' \cdot \psi = \theta \cdot \partial_z^2\psi - \delta \cdot \partial_z\psi - \partial_z(\delta \cdot \psi) \tag{18.26}$$

* * *

$$\begin{aligned}\partial_x\partial_y^2(\theta \cdot \psi) = \theta \cdot \partial_x\partial_y^2\psi + \delta \cdot \Big\{\partial_z^2\psi \cdot \partial_x z_0 \cdot (\partial_y z_0)^2 \\ +\partial_y^2\psi \cdot \partial_x z_0 + 2 \cdot \partial_y\partial_z\psi \cdot \partial_x z_0 \cdot \partial_y z_0 + \partial_x\partial_z\psi \cdot (\partial_y z_0)^2 \\ +2 \cdot \partial_x\partial_y\psi \cdot \partial_y z_0 + \partial_z\psi \cdot \left[2 \cdot \partial_y z_0 \cdot \partial_x\partial_y z_0 + \partial_x z_0 \cdot \partial_y^2 z_0\right] \\ +2 \cdot \partial_y\psi \cdot \partial_x\partial_y z_0 + \partial_x\psi \cdot \partial_y^2 z_0 + \psi \cdot \partial_x\partial_y^2 z_0\Big\} \\ - \partial_z\Big\{\delta \cdot \Big[2 \cdot \partial_z\psi \cdot \partial_x z_0 \cdot (\partial_y z_0)^2 + 2 \cdot \partial_y\psi \cdot \partial_x z_0 \cdot \partial_y z_0 + \partial_x\psi \cdot (\partial_y z_0)^2 \\ +\psi \cdot \left(2 \cdot \partial_y z_0 \cdot \partial_x\partial_y z_0 + \partial_x z_0 \cdot \partial_y^2 z_0\right)\Big]\Big\} + \partial_z^2\left[\delta \cdot \psi \cdot \partial_x z_0 \cdot (\partial_y z_0)^2\right]\end{aligned} \tag{18.27}$$

$$\partial_y^3(\theta\cdot\psi)=\theta\cdot\partial_y^3\psi+\delta\cdot\left\{\partial_z^2\psi\cdot(\partial_y z_0)^3+3\cdot\partial_y^2\psi\cdot\partial_y z_0+3\cdot\partial_y\partial_z\psi\cdot(\partial_y z_0)^2\right.$$
$$\left.+3\cdot\partial_z\psi\cdot\partial_y z_0\cdot\partial_y^2 z_0+3\cdot\partial_y\psi\cdot\partial_y^2 z_0+\psi\cdot\partial_y^3 z_0\right\}$$
$$-\partial_z\left\{\delta\cdot\left[2\cdot\partial_z\psi\cdot(\partial_y z_0)^3+3\cdot\partial_y\psi\cdot(\partial_y z_0)^2+3\cdot\psi\cdot\partial_y z_0\cdot\partial_y^2 z_0\right]\right\}$$
$$+\partial_z^2\left[\delta\cdot\psi\cdot(\partial_y z_0)^3\right] \tag{18.28}$$

18.4 Integrals

Integrals of products of a smooth function ψ and the delta function δ (18.11) or its derivative δ' (18.14) are calculated. These integrals arise from the application of integral operators.

—

$$[\cdot]_{z_0} \stackrel{\text{def.}}{=} [\cdot]_{z=z_0(x,\ y)} \tag{18.29}$$

$$[\cdot]_{y_0} \stackrel{\text{def.}}{=} [\cdot]_{y=y_0(x,\ z)} \tag{18.30}$$

$$[\cdot]_{x_0} \stackrel{\text{def.}}{=} [\cdot]_{x=x_0(y,\ z)} \tag{18.31}$$

$$\psi_{z_0}(x,y) \stackrel{\text{def.}}{=} [\psi]_{z_0}=\psi[x,y,z_0(x,\ y)] \tag{18.32}$$

$$\psi_{y_0}(x,z) \stackrel{\text{def.}}{=} [\psi]_{y_0}=\psi[x,y_0(x,\ z),z] \tag{18.33}$$

$$\psi_{x_0}(y,z) \stackrel{\text{def.}}{=} [\psi]_{x_0}=\psi[x_0(y,\ z),y,z] \tag{18.34}$$

* * *

$$\partial_z^{-1}[\delta(z_0-z)\cdot\psi]=-\theta\cdot\psi_{z_0} \tag{18.35}$$

$$\partial_y^{-1}[\delta(z_0-z)\cdot\psi]=\theta\cdot\frac{\psi_{y_0}}{[\partial_y z_0]_{y_0}} \tag{18.36}$$

$$\partial_x^{-1}[\delta(z_0 - z) \cdot \psi] = \theta \cdot \frac{\psi_{x_0}}{[\partial_x z_0]_{x_0}} \tag{18.37}$$

* * *

$$\partial_z^{-2}[\delta(z_0 - z) \cdot \psi] = (z_0 - z) \cdot \theta \cdot \psi_{z_0} \tag{18.38}$$

$$\partial_y^{-1}\partial_z^{-1}[\delta(z_0 - z) \cdot \psi] = -\partial_y^{-1}(\theta \cdot \psi_{z_0}) \tag{18.39}$$

$$\partial_x^{-1}\partial_y^{-1}[\delta(z_0 - z) \cdot \psi] = \partial_x^{-1}\left[\theta \cdot \frac{\psi_{y_0}}{[\partial_y z_0]_{y_0}}\right] = \partial_y^{-1}\left[\theta \cdot \frac{\psi_{x_0}}{[\partial_x z_0]_{x_0}}\right] \tag{18.40}$$

* * *

$$\partial_y^{-1}[\delta'(z_0 - z) \cdot \psi] \overset{(18.16)}{\underset{(18.36)}{=}} -\theta \cdot \left[\frac{\partial_y[\psi/\partial_y z_0]}{\partial_y z_0}\right]_{y_0} + \delta \cdot \frac{\psi}{\partial_y z_0} \tag{18.41}$$

$$\begin{aligned}\partial_x^{-1}\partial_y^{-1}[\delta'(z_0 - z) \cdot \psi] &\overset{(18.41)}{\underset{(18.37)}{=}} -\partial_x^{-1}\left\{\theta \cdot \left[\frac{\partial_y[\psi/\partial_y z_0]}{\partial_y z_0}\right]_{y_0}\right\} + \theta \cdot \left[\frac{\psi}{\partial_y z_0 \cdot \partial_x z_0}\right]_{x_0} \\ &= -\partial_y^{-1}\left\{\theta \cdot \left[\frac{\partial_x[\psi/\partial_x z_0]}{\partial_x z_0}\right]_{x_0}\right\} + \theta \cdot \left[\frac{\psi}{\partial_x z_0 \cdot \partial_y z_0}\right]_{y_0}\end{aligned} \tag{18.42}$$

18.5 Transformations in Model Types "a"–"e"

The following pattern transformations are used to write the summands $\mathbf{S}_*$ and $\mathbf{T}_0$ of the general solution $\mathbf{S}$ for the models of type "a"–"e"[8] as integral representations by first applying the differential operators and then the integral operators.

The typical terms of the pattern transformations arise by applying products of differential and integral operators to a product $\theta \cdot \psi$ consisting of the step function θ and of a smooth function ψ. In the considered glacier domain, this function ψ is respectively equal to one of the three independent stress components when computing $\mathbf{T}_0$ and is equal to the

[8] The summands $\mathbf{S}_*$ and $\mathbf{T}_0$ are given in Sects. 17.1, 17.2, 17.3, 17.4, and 17.5.

ice density when computing $\mathbf{S}_*$. Products of equal numbers of integral and differential operators occur in the computation of $\mathbf{T}_0$, and therefore one can apply to $\mathbf{T}_0$ the pattern transformations of such expressions in which the sum of the operator exponents vanishes (18.43)–(18.49). Operator products occur in the computation of $\mathbf{S}_*$ in which the number of integral operators is one greater than the number of differential operators. Therefore for $\mathbf{S}_*$ the pattern transformations of such expressions are decisive, in which the sum of the operator exponents is -1 (18.50)–(18.52). Here the already introduced terms for boundary values are used (18.29)–(18.34).

—

$$\partial_y^{-1}\partial_z(\theta\cdot\psi) \overset{(18.21)}{\underset{(18.36)}{=}} \partial_y^{-1}(\theta\cdot\partial_z\psi)-\theta\cdot\left[\psi/\partial_y z_0\right]_{y_0} \tag{18.43}$$

$$\partial_y^{-1}\partial_x(\theta\cdot\psi) \overset{(18.20)}{\underset{(18.36)}{=}} \partial_y^{-1}(\theta\cdot\partial_x\psi)+\theta\cdot\left[\partial_x z_0\cdot\psi/\partial_y z_0\right]_{y_0} \tag{18.44}$$

$$\partial_z^{-1}\partial_y(\theta\cdot\psi) \overset{(18.19)}{\underset{(18.35)}{=}} \partial_z^{-1}\left[\theta\cdot\partial_y\psi\right]-\theta\cdot\partial_y z_0\cdot\psi_{z_0} \tag{18.45}$$

* * *

$$\begin{aligned}\partial_x^{-1}\partial_y^{-1}\partial_z^2(\theta\cdot\psi) &\overset{(18.26)\,(18.40)\,(18.42)}{=} \partial_x^{-1}\partial_y^{-1}\left(\theta\cdot\partial_z^2\psi\right)-2\partial_y^{-1}\left[\theta\cdot\frac{\left[\partial_z\psi\right]_{x_0}}{\left[\partial_x z_0\right]_{x_0}}\right]\\ &-\partial_y^{-1}\left\{\theta\cdot\left[\frac{\partial_x\left[\psi/\partial_x z_0\right]}{\partial_x z_0}\right]_{x_0}\right\}+\theta\cdot\left[\frac{\psi}{\partial_x z_0\cdot\partial_y z_0}\right]_{y_0}\end{aligned} \tag{18.46}$$

$$\partial_z^{-1}\partial_y^{-1}\partial_x^2(\theta\cdot\psi) \overset{(18.24)\,(18.39)\,(18.36)}{=} \partial_z^{-1}\partial_y^{-1}\left[\theta\cdot\partial_x^2\psi\right] - \partial_y^{-1}\left\{\theta\cdot\left[2\cdot\partial_x z_0\cdot\partial_x\psi + (\partial_x z_0)^2\cdot\partial_z\psi + \partial_x^2 z_0\cdot\psi\right]_{z_0}\right\} - \theta\cdot\left[(\partial_x z_0)^2\cdot(\partial_y z_0)^{-1}\cdot\psi\right]_{y_0} \tag{18.47}$$

$$\partial_z^{-2}\partial_y^2(\theta\cdot\psi) \overset{(18.23)\,(18.35)\,(18.38)}{=} \partial_z^{-2}\left[\theta\cdot\partial_y^2\psi\right] + (z_0 - z)\cdot\theta \cdot\left[2\cdot\partial_y z_0\cdot\partial_y\psi + \left(\partial_y z_0\right)^2\cdot\partial_z\psi + \partial_y^2 z_0\cdot\psi\right]_{z_0} + \theta \cdot\left(\partial_y z_0\right)^2\cdot\psi_{z_0} \tag{18.48}$$

$$\partial_z^{-2}\partial_x\partial_y(\theta\cdot\psi) \overset{(18.22)\,(18.35)\,(18.38)}{=} \partial_z^{-2}\left[\theta\cdot\partial_x\partial_y\psi\right] + (z_0 - z)\cdot\theta\cdot \cdot\left[\partial_x z_0\cdot\partial_y\psi + \partial_y z_0\cdot\partial_x\psi + \partial_x z_0\cdot\partial_y z_0\cdot\partial_z\psi + \partial_x\partial_y z_0\cdot\psi\right]_{z_0} + \theta\cdot\partial_x z_0\cdot\partial_y z_0\cdot\psi_{z_0} \tag{18.49}$$

* * *

$$\partial_x^{-1}\partial_y^{-1}\partial_z(\theta\cdot\psi) \overset{(18.43)}{=} \partial_x^{-1}\partial_y^{-1}(\theta\cdot\partial_z\psi) - \partial_x^{-1}\left\{\theta\cdot\left[\psi/\partial_y z_0\right]_{y_0}\right\} = \partial_x^{-1}\partial_y^{-1}(\theta\cdot\partial_z\psi) - \partial_y^{-1}\left\{\theta\cdot\left[\psi/\partial_x z_0\right]_{x_0}\right\} \tag{18.50}$$

$$\partial_z^{-2}\partial_y(\theta\cdot\psi) \overset{(18.19)\,(18.38)}{=} \partial_z^{-2}\left(\theta\cdot\partial_y\psi\right) + (z_0 - z)\cdot\theta\cdot\partial_y z_0\cdot\psi_{z_0} \tag{18.51}$$

$$\partial_y^{-1}\partial_z^{-1}\partial_x(\theta\cdot\psi) \overset{(18.20)\,(18.39)}{=} \partial_y^{-1}\partial_z^{-1}(\theta\cdot\partial_x\psi) - \partial_y^{-1}\left\{\theta\cdot\partial_x z_0\cdot\psi_{z_0}\right\} \tag{18.52}$$

18.6 Transformations in the Model Types "f"–"g"

Only the model type "f" in Sect. 17.6 is examined with the hyperbolic differential operator $\Box_z^{-1}$. The model type "g" in Sect. 17.7 contains the hyperbolic differential operator $\Box_y^{-1}$ and the calculations with this operator are analogous to the calculations with $\Box_z^{-1}$. In the following, we calculate the typical terms that make up the summands $\mathbf{T}_0$ (17.48) and $\mathbf{S}_f$ (17.49) of the general solution $\mathbf{S}$ in model type "f".

These typical terms arise by applying products of differential and integral operators to a product $\theta \cdot \psi$ consisting of the step function θ (18.10) and a smooth function ψ. This function ψ is equal to each of the three independent deviatoric stress components in the glacier domain when computing $\mathbf{T}_0$ and is equal to the ice density when computing $\mathbf{S}_f$. The calculation of the typical terms is done in two steps. The terms already introduced for boundary values (18.29)–(18.34) are used.

In the first step, all differential operators are applied first, and then all integral operators are applied except the integral operator $\Box_z^{-1}$. The result in each case is a distribution, (18.53) in which the step function θ (18.10), the delta function δ (18.11) and the smooth functions q_1, q_2 and q_3 occur. The typical expressions of this first step are tabulated below. Their distributional forms (18.53) are in the formulas with the numbers given. Occurring expressions of the type $\partial_z^{-1}(\theta \cdot \psi)$ in (18.54)–(18.57) can be written as $\theta \cdot \partial_z^{-1}(\theta \cdot \psi)$ and therefore also match this form (18.53).

formula number	typical expression	distributional form
(18.21)	$\partial_z(\theta \cdot \psi)$	
(18.19)	$\partial_y(\theta \cdot \psi)$	
(18.54)	$\partial_z^{-1}\partial_y^2(\theta \cdot \psi)$	
(18.55)	$\partial_z^{-1}\partial_x\partial_y(\theta \cdot \psi)$	$= \theta \cdot q_1(x, y, z)$
(18.26)	$\partial_z^2(\theta \cdot \psi)$	$+\delta \cdot q_2(x, y)$
(18.23)	$\partial_y^2(\theta \cdot \psi)$	$+\partial_z[\delta \cdot q_3(x, y)]$
(18.25)	$\partial_z\partial_y(\theta \cdot \psi)$	
(18.22)	$\partial_x\partial_y(\theta \cdot \psi)$	
(18.56)	$\partial_z^{-1}\partial_y^2\partial_x(\theta \cdot \psi)$	
(18.57)	$\partial_z^{-1}\partial_y^3(\theta \cdot \psi)$	

The typical expressions where the sum of the operator exponents is 1, i.e. the first four expressions in the list, are used to compute the summand $\mathbf{S}_f$ (17.49) of the general solution, and the typical expressions where the sum of the operator exponents is 2 are used to compute the summand $\mathbf{T}_0$ (17.48).

In the second step the operator $\Box_z^{-1}$ is applied to distributions of the mentioned type (18.58). All matrix elements of $\mathbf{T}_0$ (17.48) and $\mathbf{S}_f$ (17.49) can be written in this form

(18.58).[9] In general, the z-derivative of an integral also occurs. This term arises from an integral representation with a distribution as integrand, but cannot be transformed into an integral with an ordinary function as integrand, thus does not admit an ordinary integral representation. The matrix elements of $\mathbf{S}_f$ do not contain such a term with z-derivative, but the matrix elements of $\mathbf{T}_0$ do.

—

$$\theta \cdot q_1(x, y, z) + \delta \cdot q_2(x, y) + \partial_z[\delta \cdot q_3(x, y)] \tag{18.53}$$

* * *

$$\partial_z^{-1}\partial_y^2(\theta \cdot \psi) \overset{(18.23)}{=} \partial_z^{-1}\left[\theta \cdot \partial_y^2\psi\right] - \theta \cdot \left[\partial_z\psi \cdot (\partial_y z_0)^2 + 2 \cdot \partial_y\psi \cdot \partial_y z_0 + \psi \cdot \partial_y^2 z_0\right]_{z_0} - \delta \cdot \psi \cdot (\partial_y z_0)^2 \tag{18.54}$$

$$\partial_z^{-1}\partial_x\partial_y(\theta \cdot \psi) \overset{(18.22)}{=} \partial_z^{-1}(\theta \cdot \partial_x\partial_y\psi) - \theta \cdot \left[\partial_z\psi \cdot \partial_x z_0 \cdot \partial_y z_0 + \partial_x\psi \cdot \partial_y z_0 + \partial_y\psi \cdot \partial_x z_0 + \psi \cdot \partial_x\partial_y z_0\right]_{z_0} - \delta \cdot \psi \cdot \partial_x z_0 \cdot \partial_y z_0 \tag{18.55}$$

$$\begin{aligned}\partial_z^{-1}\partial_x\partial_y^2(\theta \cdot \psi) \overset{(18.27)}{=} {} & \partial_z^{-1}\left(\theta \cdot \partial_x\partial_y^2\psi\right) - \theta \cdot \left[\partial_z^2\psi \cdot \partial_x z_0 \cdot (\partial_y z_0)^2 + \partial_y^2\psi \cdot \partial_x z_0\right. \\ & +2 \cdot \partial_y\partial_z\psi \cdot \partial_x z_0 \cdot \partial_y z_0 + \partial_x\partial_z\psi \cdot (\partial_y z_0)^2 + 2 \cdot \partial_x\partial_y\psi \\ & \cdot \partial_y z_0 + \partial_z\psi \cdot \left(2 \cdot \partial_y z_0 \cdot \partial_x\partial_y z_0 + \partial_x z_0 \cdot \partial_y^2 z_0\right) + 2 \\ & \left.\cdot \partial_y\psi \cdot \partial_x\partial_y z_0 + \partial_x\psi \cdot \partial_y^2 z_0 + \psi \cdot \partial_x\partial_y^2 z_0\right]_{z_0} - \delta \\ & \cdot\left[2 \cdot \partial_z\psi \cdot \partial_x z_0 \cdot (\partial_y z_0)^2\right. \\ & +2 \cdot \partial_y\psi \cdot \partial_x z_0 \cdot \partial_y z_0 + \partial_x\psi \cdot (\partial_y z_0)^2 \\ & \left.+\psi \cdot \left(2 \cdot \partial_y z_0 \cdot \partial_x\partial_y z_0 + \partial_x z_0 \cdot \partial_y^2 z_0\right)\right] \\ & +\partial_z\left[\delta \cdot \psi \cdot \partial_x z_0 \cdot (\partial_y z_0)^2\right]\end{aligned} \tag{18.56}$$

[9] The summands in (18.58) are each solutions of a classical hyperbolic differential equation in three variables with boundary conditions and are discussed in Sect. 19. These boundary conditions of the classical problem are used to calculate the stress tensor fields at the free surface Σ (10.54). See footnote 27 in Chapter 10.

$$\partial_z^{-1}\partial_y^3(\theta\cdot\psi) \overset{(18.28)}{=} \partial_z^{-1}\left(\theta\cdot\partial_y^3\psi\right) - \theta$$
$$\cdot\left[\partial_z^2\psi\cdot(\partial_y z_0)^3 + 3\cdot\partial_y^2\psi\cdot\partial_y z_0 + 3\cdot\partial_y\partial_z\psi\cdot(\partial_y z_0)^2\right.$$
$$\left.+3\cdot\partial_z\psi\cdot\partial_y z_0\cdot\partial_y^2 z_0 + 3\cdot\partial_y\psi\cdot\partial_y^2 z_0 + \psi\cdot\partial_y^3 z_0\right]_{z0} - \delta$$
$$\cdot\left[2\cdot\partial_z\psi\cdot(\partial_y z_0)^3 + 3\cdot\partial_y\psi\cdot(\partial_y z_0)^2\right.$$
$$\left.+3\cdot\psi\cdot\partial_y z_0\cdot\partial_y^2 z_0\right] + \partial_z\left[\delta\cdot\psi\cdot(\partial_y z_0)^3\right] \tag{18.57}$$

* * *

$$\square_z^{-1}[\theta\cdot q_1(x,y,z) + \delta\cdot q_2(x,y) + \partial_z[\delta\cdot q_3(x,\ y)]$$
$$= \int dx'dy'dz'\cdot G(\boldsymbol{r}'-\boldsymbol{r})\cdot\theta(\boldsymbol{r}')\cdot q_1(\boldsymbol{r}') + \int dx'dy'\cdot[G(\boldsymbol{r}'-\boldsymbol{r})]_{\prime z_0}$$
$$\cdot q_2(x',y') + \partial_z\int dx'dy'\cdot[G(\boldsymbol{r}'-\boldsymbol{r})]_{\prime z_0}\cdot q_3(x',y') \tag{18.58}$$

$$\theta(\boldsymbol{r}')\overset{\text{def}}{=}\theta[z_0(x',y')-z'];\quad [\cdot]_{\prime z_0}\overset{\text{def}}{=}[\cdot]_{z'=z_0(x',y')}.$$

18.7 Transformations in Model Type "h"

The transformations of the general solution **S** into integral representations can be performed for the models of type "h"[10] in a similar way as for the models of type "a"–"e". To do this, one introduces a skew η-ξ coordinate system in the x-z plane and converts everything to the skew η-ξ-y coordinate system (18.59)–(18.65).[11]

[10]The summands $\mathbf{S}_*$ and $\mathbf{T}_0$ are given in Sect. 17.8.

[11]The conversion to the new η-ξ-y locational coordinates serves only for a simpler representation of the locational dependence. The tensor components of the stress tensors are not transformed and still refer to the original Cartesian x-y-z coordinate system. A conversion of the tensor components to the oblique system would not bring any simplification and the symmetry of the stress tensors would be lost.

—

$$\mathbf{e}_\eta = \frac{1}{\sqrt{3}}\left(\mathbf{e}_x + \sqrt{2}\cdot\mathbf{e}_z\right); \mathbf{e}_\xi = \frac{1}{\sqrt{3}}\left(-\mathbf{e}_x + \sqrt{2}\cdot\mathbf{e}_z\right) \quad (18.59)$$

$$\mathbf{e}_x = \frac{\sqrt{3}}{2}(\mathbf{e}_\eta - \mathbf{e}_\xi); \mathbf{e}_z = \frac{\sqrt{3}}{2\sqrt{2}}(\mathbf{e}_\eta + \mathbf{e}_\xi) \quad (18.60)$$

$$x = \frac{1}{\sqrt{3}}(\eta - \xi); z = \frac{\sqrt{2}}{\sqrt{3}}(\eta + \xi) \quad (18.61)$$

$$\eta = \frac{\sqrt{3}}{2\sqrt{2}}\left(\sqrt{2}x + z\right); \xi = \frac{\sqrt{3}}{2\sqrt{2}}\left(-\sqrt{2}x + z\right) \quad (18.62)$$

$$\partial_x = \frac{\sqrt{3}}{2}(\partial_\eta - \partial_\xi); \partial_z = \frac{\sqrt{3}}{2\sqrt{2}}(\partial_\eta + \partial_\xi) \quad (18.63)$$

$$\partial_\eta = \frac{1}{\sqrt{3}}\left(\partial_x + \sqrt{2}\partial_z\right); \partial_\xi = \frac{1}{\sqrt{3}}\left(-\partial_x + \sqrt{2}\partial_z\right) \quad (18.64)$$

$$\left(\partial_x^2 - 2\partial_z^2\right) = -3\partial_\eta\partial_\xi; \left(\partial_x^2 - 2\partial_z^2\right)^{-1} = -\frac{1}{3}\cdot\partial_\eta^{-1}\partial_\xi^{-1} \quad (18.65)$$

The Hyperbolic Differential Equation in Three Variables

19

Abstract

The solution of the hyperbolic differential equation given in distributional form is explained. This solution occurs in the models "f" and "g", in each of which three selected components of the deviatoric stress tensor can be taken as arbitrary functions within the framework of the general solution.

The solution of the hyperbolic differential equation given in distributional form plays a role in models of types "f" and "g".[1]

For calculations in distributional form, the spatial domain of definition includes not only the glacier domain Ω, but also an external domain Ω_{ext} beyond the free surface Σ (19.1) where all admissible functions and distributions vanish.[2] The right-hand side of the hyperbolic differential equation (19.12) is a distribution in which the step function θ (19.5), the delta function δ (19.6), and the smooth functions q_1, q_2, and q_3 occur, which are known. The solution $\theta \cdot \chi$ of this differential equation is a smooth function χ in the glacier domain and vanishes in the external domain.[3] This solution $\theta \cdot \chi$ (19.13) arises in distributional form by applying the inverse hyperbolic operator $\Box_z^{-1}$ (19.2) to the distribution, which is on the right-hand side of the differential equation.[4]

These distributional forms (19.12) and (19.13) of the hyperbolic differential equation and its solution can also be interpreted classically. The hyperbolic differential equation in

[1] See paragraphs 17.6, 17.7, 18.6.

[2] See Sect. 18.1.

[3] The function $z_0(x, y)$ (19.1) should also be smooth.

[4] See Sect. 3.5.

P. Halfar, *Stresses in glaciers*, https://doi.org/10.1007/978-3-662-66024-9_19

distributional form corresponds to a classical boundary value problem and the solution given in distributional form is an ordinary function in the glacier domain Ω.

To define the corresponding classical boundary value problem, one mathematically transforms the left-hand side of the distributional differential equation (19.12), expressing the x and y derivatives of the smooth function χ on the surface Σ by its z-derivative and by the x and y derivatives of its boundary values χ_{z_0}, respectively, using the differentiation rules (19.9) and (19.10). A comparison of this transformation (19.11) with the right-hand side of this differential equation (19.12) shows that the solution $\theta \cdot \chi$ (19.13) of this differential equation solves the following classical boundary value problem (19.14):

1. In the glacier domain, the solution $\theta \cdot \chi$ satisfies the inhomogeneous hyperbolic differential equation with the right-hand side q_1.
2. The boundary value χ_{z_0} of the solution at the free surface Σ is defined by the function q_3.
3. The boundary value $[\partial_z \chi]_{z_0}$ of the z-derivative is defined by the functions q_2 and q_3 and also by the derivatives of q_3.[5]

The solution $\theta \cdot \chi$ given in distributional form (19.13) is a sum of three ordinary functions. The value of the first function (19.15) at a point $\mathbf{r}$ is obtained by integration over the dependence cone starting from the point $\mathbf{r}$,[6] with no contributions coming from the cone region beyond the free surface Σ, since the integrand vanishes there. This function (19.15) is a solution of the boundary value problem (19.14) with vanishing functions q_2 and q_3, so it solves the inhomogeneous differential equation with the right-hand side q_1 and vanishes together with its z-derivative on the boundary surface Σ.

The value of the second function (19.16) at a point $\mathbf{r}$ is obtained by integration over that part of the free surface Σ which lies in this dependence cone. This function (19.16) is a solution of the boundary value problem (19.14) with vanishing functions q_1 and q_3, so it solves the homogeneous differential equation, vanishes on the boundary surface Σ and has there a z-derivative defined by q_2.

The value of the third function (19.17) cannot be written as an integral with an ordinary function in the integrand, but as the z-derivative of a function which is of the same type as the second function. This function (19.17) is a solution of the boundary value problem (19.14) with vanishing functions q_1 and q_2, so it solves the homogeneous differential equation, and its boundary values and the boundary values of its z-derivative on the boundary surface Σ are defined by q_3.

[5] The dependence of q_3 arises by expressing the boundary function χ_{z_0} according to no. 2 by q_3.

[6] The function $G(\mathbf{r}' - \mathbf{r})$ (19.3) is different from zero only in the points $\mathbf{r}'$ which lie in the dependence cone starting from the point $\mathbf{r}$.

—

$$\Sigma: \quad z = z_0(x, y) \tag{19.1}$$

$$\Box_z^{-1}\chi \overset{(3.19)}{=} \int dx'\,dy'\,dz' \cdot G(\mathbf{r}' - \mathbf{r}) \cdot \chi(\mathbf{r}') \tag{19.2}$$

$$G(\mathbf{r}' - \mathbf{r}) \overset{(3.20)}{=} \frac{1}{2\pi} \cdot \frac{\theta\left[(z' - z) - \sqrt{(x' - x)^2 + (y' - y)^2}\right]}{\sqrt{(z' - z)^2 - (x' - x)^2 - (y' - y)^2}} \tag{19.3}$$

$$\mathbf{r} = (x, y, z)^{\mathrm{T}}; \quad \mathbf{r}' = (x', y', z')^{\mathrm{T}} \tag{19.4}$$

$$\theta \overset{\text{def.}}{=} \theta[z_0(x, y) - z] \tag{19.5}$$

$$\delta \overset{\text{def.}}{=} \delta[z_0(x, y) - z] \tag{19.6}$$

$$[\cdot]_{z_0} \overset{\text{def.}}{=} [\cdot]_{z = z_0(x,y)} \tag{19.7}$$

$$\chi_{z_0}(x, y) \overset{\text{def.}}{=} \chi_{z_0} = \chi[x, y, z_0(x, y)] \tag{19.8}$$

* * *

$$\partial_x \chi_{z_0} \overset{\text{id.}}{=} [\partial_x \chi]_{z_0} + [\partial_z \chi]_{z_0} \cdot \partial_x z_0 \tag{19.9}$$

$$\partial_y \chi_{z_0} \overset{\text{id.}}{=} [\partial_y \chi]_{z_0} + [\partial_z \chi]_{z_0} \cdot \partial_y z_0 \tag{19.10}$$

$$\begin{aligned}\Box_z(\theta \cdot \chi) &\overset{\text{id.}}{=} \theta \cdot \Box_z \chi - \delta \cdot \Big\{\left(\partial_x^2 z_0 + \partial_y^2 z_0\right) \cdot \chi_{z_0} + 2\left(\partial_x z_0 \cdot \partial_x \chi_{z_0} + \partial_y z_0 \cdot \partial_y \chi_{z_0}\right) \\ &+ [\partial_z \chi]_{z_0} \cdot \left[1 - (\partial_x z_0)^2 - (\partial_y z_0)^2\right]\Big\} - \partial_z\left\{\delta \cdot \chi_{z_0} \cdot \left[1 - (\partial_x z_0)^2 - (\partial_y z_0)^2\right]\right\}\end{aligned} \tag{19.11}$$

* * *

$$\Box_z(\theta \cdot \chi) = \theta \cdot q_1(x,y,z) + \delta \cdot q_2(x,y) + \partial_z[\delta \cdot q_3(x,y)] \Updownarrow \tag{19.12}$$

$$\theta \cdot \chi = \Box_z^{-1}\{\theta \cdot q_1(x,y,z) + \delta \cdot q_2(x,y) + \partial_z[\delta \cdot q_3(x,y)]\} \Updownarrow \tag{19.13}$$

$$\left.\begin{array}{rcl} \theta\cdot\Box_z\chi & = & \theta\cdot q_1 \\ {[\partial_z\chi]}_{z_0} & = & \left[-q_2-\left(\partial_x^2 z_0+\partial_y^2 z_0\right)\cdot\chi_{z_0}\ -2\left(\partial_x z_0\cdot\partial_x\chi_{z_0}+\partial_y z_0\cdot\partial_y\chi_{z_0}\right)\right] \\ & & \cdot\left[1-(\partial_x z_0)^2-\left(\partial_y z_0\right)^2\right]^{-1} \\ \chi_{z_0} & = & -q_3\cdot\left[1-(\partial_x z_0)^2-\left(\partial_y z_0\right)^2\right]^{-1} \end{array}\right\} \tag{19.14}$$

* * *

$$\begin{aligned} &[\Box_z^{-1}(\theta\cdot q_1)](\mathbf{r})=\int \mathrm{d}x'\,\mathrm{d}y'\,\mathrm{d}z'\cdot G(\mathbf{r}'-\mathbf{r})\cdot\theta(\mathbf{r}')\cdot q_1(\mathbf{r}'); \\ &\theta(\mathbf{r}')\overset{\text{def.}}{=}\theta[z_0(x',y')-z'] \end{aligned} \tag{19.15}$$

$$[\Box_z^{-1}(\delta\cdot q_2)](\mathbf{r})=\int \mathrm{d}x'\,\mathrm{d}y'\cdot[G(\mathbf{r}'-\mathbf{r})]_{'z_0}\cdot q_2(x',y');\quad [\cdot]_{'z_0}\overset{\text{def}}{=}[\cdot]_{z'=z_0(x',y')} \tag{19.16}$$

$$[\Box_z^{-1}\partial_z(\delta\cdot q_3)](\mathbf{r})=[\partial_z\Box_z^{-1}(\delta\cdot q_3)](\mathbf{r})=\partial_z\int \mathrm{d}x'\mathrm{d}y'\cdot[G(\mathbf{r}'-\mathbf{r})]_{'z_0}\cdot q_3(x',y') \tag{19.17}$$

Table Icebergs

20

Abstract

Details of the horizontally isotropic homogeneous, infinitely extended tabular iceberg models are discussed.

20.1 The Functions K_1, K_2, χ, I_1 and I_2

In the following some properties of the functions $K_1(d), K_2(d), \chi(C_1), I_1(\lambda_*, \Delta)$ and $I_2(\lambda_*, \Delta)$ (20.5)–(20.9) are presented (Cf. Fig. 11.1). In the case of the functions I_1 and I_2 the area in the λ_*-Δ – coordinate system, which lies between the ordinate and its right parallel at a distance of 1, plays a role, which in the following is called "strip" (20.10). The areas of the λ_*-Δ – coordinate system which lie above and below the abscissa and are marked by positive and negative values of Δ, respectively, are designated as the "upper half-plane" and "lower half-plane".

1. Monotonicity, zeros and ranges of values of the functions K_1, K_2 and χ

   The functions K_1, K_2, and χ (20.7)–(20.9) are each monotonically increasing, vanishing for vanishing argument, and their range of values extends from minus to plus infinity.

   Justification: That K_1 and K_2 are monotonically increasing and vanish for vanishing argument follows from the corresponding properties (20.1) and (20.2) of the flow law function Φ. That their range of values extends from minus to plus infinity (20.11) and (20.12) follows from the uniformly unrestricted growth of the flow law function Φ (20.3) and (20.4). The inverse functions of K_1 and K_2 also have these properties, so the function χ also has them.

P. Halfar, *Stresses in glaciers*, https://doi.org/10.1007/978-3-662-66024-9_20

2. Sign of the functions I_1 and I_2 Outside the strip

   The signs of the functions I_1 and I_2 are positive to the left of the strip and negative to the right of the strip in the upper half-plane, but negative to the left of the strip and positive to the right of the strip in the lower half-plane.

   Justification: These signs of functions I_1 and I_2 result from the zero and monotonicity of the flow law function Φ (20.1) and (20.2).
3. Monotonicity and value range of the functions I_1 and I_2 as functions of λ_*

   The functions I_1 and I_2 are monotonically decreasing in λ_* in the upper half plane and monotonically increasing in the lower half plane. Their values traverse all real numbers if λ_* traverses all real numbers.

   Justification: The monotonicity properties of functions I_1 and I_2 follow from the monotonicity (20.2) of the flow law function Φ. That the values of these functions pass through all real numbers follows from the behavior of these functions with unrestricted increasing or decreasing λ_* (20.13) and (20.14).
4. Monotonicity and range of values of functions I_1 and I_2 as functions of Δ Outside the strip

   The functions I_1 and I_2 are monotone functions of Δ outside the strip. These functions are monotonically increasing in Δ to the left of the strip and monotonically decreasing in Δ to the right of the strip. Their values traverse all real numbers except zero if Δ traverses all real numbers except zero.

   Justification: The monotonicity properties of functions I_1 and I_2 follow from the monotonicity (20.2) of the flow law function Φ. That the values of these functions pass through all real numbers except zero follows from the behavior of these functions when Δ is unboundedly increasing or decreasing (20.13) and (20.14), and from their convergence to zero when Δ converging to zero.
5. The zero level lines of the functions I_1 and I_2

   The zero level lines of functions I_1 and I_2 lie in the strip and are each a function of Δ.[1] The zero level line of I_2 lies to the right of the zero level line of I_1.

   Justification: That the zero-level lines of I_1 and I_2 must be functions of Δ follows from the monotonicity properties (no. 3) of functions I_1 and I_2. That they lie in the strip follows from the different signs of the function values of I_1 and I_2, respectively, to the left and right of the strip (no. 2). The zero level line of I_2 must lie to the right of the zero level line of I_1, because the function I_2 on the zero level line (20.20) of I_1 has a sign opposite to the sign on the right side of the strip (20.21).
6. The level lines of the function I_1
   (a) The level lines of I_1 as functions of Δ

       For every real I_1 value C_1 there is a level line and this level line is a function $\lambda_{C_1}(\Delta)$ (20.19) of Δ, which is defined for all nonvanishing Δ.

[1] "Function of Δ" means that for every value Δ there is exactly one matching value λ_*.

Justification: This follows from the monotonicity and the range of values of the function I_1 as a function of λ_* (no. 3).

(b) The level lines of function I_1 outside the strip as functions of λ_*

Each level line of the function I_1, except the zero level line, is a function of λ_*, defined for all λ_* outside the strip.

Justification: This follows from the monotonicity and the range of values of the function I_1 as a function of Δ (no. 4).

(c) Position of level lines of I_1 to non-vanishing I_1 values C_1

The level lines (20.19) of the function I_1 to non-vanishing I_1 values C_1 consist of two branches. Each branch lies either in the upper half-plane or in the lower half-plane (in short: above or below) and it lies either to the left or to the right of the zero level line of I_1 (in short: left or right). The two branches of a level line are always diagonally opposite each other: For negative I_1 value, one branch is below left and the other branch is above right; for positive I_1 value, one branch is above left and the other branch is below right.

Justification: This follows from the monotonicity of I_1 in λ_* (no. 3).

(d) Direction of the level lines and of the gradients of I_1 outside the strip

The level lines are always directed towards one coordinate axis and away from the other coordinate axis. The gradients to the level lines above left and below right are directed away from both coordinate axes and the gradients to the level lines below left and above right are directed towards both coordinate axes.

Justification: This follows from the signs of the derivatives (20.15) and (20.16) of I_1.

(e) Asymptote of the level lines from I_1 to non-vanishing I_1 values C_1

The level lines (20.19) of the function I_1 to non-vanishing I_1 values C_1 have the abscissa as asymptote, and pass into hyperbolas as this asymptote is approached (20.23).

Justification: When approaching the abscissa on the level line, the expression for the constant C_1 (20.22) keeps the product of the two coordinates Δ and $\lambda_{C_1}(\Delta)$ bounded because the flow law function Φ has uniform asymptotic properties (20.3) and (20.4). Therefore, everything takes place in a compact domain within the two-dimensional domain of definition of the flow law function Φ. In this compact domain, this function Φ is uniformly continuous, which is why one can interchange the boundary transition with integration in the expression for the constant C_1 (20.22). It follows that the level line of I_1 to the value C_1 asymptotically transitions to a hyperbola as the product of its two coordinates Δ and $\lambda_{C_1}(\Delta)$ converges to a nonvanishing limit (20.23).

7. The values of the function I_2 on the level lines of I_1

(a) Monotonicity of the function I_2

The value of the function I_2 increases when moving from the bottom to the top in the λ_*-Δ – coordinate system on an I_1 level line, i.e. when Δ is increasing.

Justification: The derivatives of the function I_2 according to the arc length of the I_1 level line in the direction of increasing Δ-values are positive (20.27). Here the directions of derivatives in the lower half-plane are defined by the tangential vectors **t** (20.25) of the I_1 level line, which arise by rotating the gradient vectors by a right angle in each case. In contrast, the directions of derivative in the upper half-plane are defined by the negative tangential vectors $-\mathbf{t}$.

(b) Excluded values of function I_2

For each I_1 level line,[2] there is an excluded I_2 value that the I_2 function does not take on that level line and that is given by the function $\chi(C_1)$ of the I_1 value C_1. The I_2 function value converges to this excluded I_2 value as one approaches the abscissa on the I_1 level line (20.28).

Justification: That this I_2 value (20.28) is excluded follows from the monotonicity of the I_2 function on the I_1 level line.

(c) Range of values of function I_2

The range of values of the I_2 function on the I_1 level line contains all values except the excluded I_2 value (20.28).

Justification: This follows from the monotonicity of the function I_2 on the I_1 level line, from the excluded I_2 value, and from the behavior of the function I_2 as Δ grows or decreases unrestrictedly (20.30) and (20.31).

$$\Phi(\lambda, 0) \overset{\text{prec.}}{=} 0 \tag{20.1}$$

$$\frac{\Phi(\lambda, d') - \Phi(\lambda, d)}{d' - d} \overset{\text{prec.}}{>} 0; \quad d' \neq d \tag{20.2}$$

$$\check{\Phi}(d) \mid \overset{\text{prec.}}{\leq} \mid \Phi(\lambda, d) \mid \tag{20.3}$$

$$d \to \pm\infty: \quad \mid \check{\Phi}(d) \mid \overset{\text{prec.}}{\to} \infty \tag{20.4}$$

* * *

[2] Only the upper and lower half planes of the λ_*-Δ coordinate system are considered, so the abscissa $\Delta = 0$ is left out.

$$I_1(\lambda_*, \Delta) \overset{\text{def}}{=} \int_0^1 \mathrm{d}\lambda \cdot \Phi[\lambda, \Delta(\lambda - \lambda_*)] \tag{20.5}$$

$$I_2(\lambda_*, \Delta) \overset{\text{def}}{=} \int_0^1 \mathrm{d}\lambda \cdot \lambda \cdot \Phi[\lambda, \Delta(\lambda - \lambda_*)] \tag{20.6}$$

$$K_1(d) \overset{\text{def}}{=} \int_0^1 \mathrm{d}\lambda \cdot \Phi(\lambda, d) \tag{20.7}$$

$$K_2(d) \overset{\text{def}}{=} \int_0^1 \mathrm{d}\lambda \cdot \lambda \cdot \Phi(\lambda, d) \tag{20.8}$$

$$\chi(C_1) \overset{\text{def.}}{=} K_2\left[\overset{-1}{K_1}(C_1)\right] \tag{20.9}$$

* * *

$$\text{strip} = \{(\lambda_*,\ \Delta) | 0 < \lambda_* < 1\} \tag{20.10}$$

* * *

$$d \to \pm\infty :$$

$$|\, K_1(d) \,| = \left| \int_0^1 \mathrm{d}\lambda \cdot \Phi(\lambda, d) \right| = \int_0^1 \mathrm{d}\lambda \cdot |\, \Phi(\lambda, d) \,| \geq \int_0^1 \mathrm{d}\lambda \cdot |\, \check{\Phi}(d) \,| \overset{(20.4)}{\rightarrow} \infty \tag{20.11}$$

$$|\, K_2(d) \,| = \left| \int_0^1 \mathrm{d}\lambda \cdot \lambda \cdot \Phi(\lambda, d) \right| = \int_0^1 \mathrm{d}\lambda \cdot \lambda \cdot |\, \Phi(\lambda, d) \,| \geq \int_0^1 \mathrm{d}\lambda \cdot \lambda \cdot |\, \check{\Phi}(d) \,| \overset{(20.4)}{\rightarrow} \infty \tag{20.12}$$

* * *

$$\Delta \neq 0; \quad \lambda_* \notin (0,1); \quad \lambda_* \to \pm\infty \quad \text{or} \quad \Delta \to \pm\infty :$$

$$\begin{aligned} |\, I_1(\lambda_*, \Delta) \,| &= \left| \int_0^1 \mathrm{d}\lambda \cdot \Phi[\lambda, \Delta(\lambda - \lambda_*)] \right| = \int_0^1 \mathrm{d}\lambda \cdot |\, \Phi[\lambda, \Delta(\lambda - \lambda_*)] \,| \\ &\geq \int_0^1 \mathrm{d}\lambda \cdot |\, \check{\Phi}\, [\Delta(\lambda - \lambda_*)] \,| \overset{(20.4)}{\rightarrow} \infty \end{aligned} \tag{20.13}$$

$$\begin{aligned}| I_2(\lambda_*,\Delta) |&= \left|\int_0^1 \mathrm{d}\lambda \cdot \lambda \cdot \Phi[\lambda,\Delta(\lambda-\lambda_*)]\right| = \int_0^1 \mathrm{d}\lambda \cdot \lambda \cdot | \Phi[\lambda,\Delta(\lambda-\lambda_*)] | \\ &\geq \int_0^1 \mathrm{d}\lambda \cdot \lambda \cdot | \check{\Phi}[\Delta(\lambda-\lambda_*)] | \overset{(20.4)}{\rightarrow} \infty \end{aligned} \tag{20.14}$$

* * *

$$\partial_{\lambda_*} I_1(\lambda_*, \Delta) = -\Delta \cdot \int_0^1 \mathrm{d}\lambda \cdot \partial_d \Phi[\lambda, \Delta(\lambda-\lambda_*)] \tag{20.15}$$

$$\partial_\Delta I_1(\lambda_*, \Delta) = \int_0^1 \mathrm{d}\lambda \cdot (\lambda-\lambda_*) \cdot \partial_d \Phi[\lambda, \Delta(\lambda-\lambda_*)] \tag{20.16}$$

$$\partial_{\lambda_*} I_2(\lambda_*, \Delta) = -\Delta \cdot \int_0^1 \mathrm{d}\lambda \cdot \lambda \cdot \partial_d \Phi[\lambda, \Delta(\lambda-\lambda_*)] \tag{20.17}$$

$$\partial_\Delta I_2(\lambda_*, \Delta) = \int_0^1 \mathrm{d}\lambda \cdot \lambda \cdot (\lambda-\lambda_*) \cdot \partial_d \Phi[\lambda, \Delta(\lambda-\lambda_*)] \tag{20.18}$$

* * *

$$\lambda_* = \lambda_{C_1}(\Delta); I_1[\lambda_{C_1}(\Delta), \Delta] = C_1 \tag{20.19}$$

$$\lambda_* = \lambda_0(\Delta); I_1[\lambda_0(\Delta), \Delta] = 0 \tag{20.20}$$

$$I_2(\lambda_0, \Delta) = \int_0^1 \mathrm{d}\lambda \cdot \underbrace{(\lambda-\lambda_0) \cdot \Phi[\lambda, \Delta(\lambda-\lambda_0)]}_{\mathrm{sign}=\mathrm{sign}(\Delta)} \tag{20.21}$$

* * *

$$\begin{aligned}C_1 &= \int_0^1 \mathrm{d}\lambda \cdot \Phi\{\lambda, \Delta[\lambda-\lambda_{C_1}(\Delta)]\} = \lim_{\Delta\to 0} \int_0^1 \mathrm{d}\lambda \cdot \Phi\{\lambda, \Delta[\lambda-\lambda_{C_1}(\Delta)]\} \\ &= K_1\left[-\lim_{\Delta\to 0} \Delta \cdot \lambda_{C_1}(\Delta)\right]\end{aligned} \tag{20.22}$$

$$\lim_{\Delta\to 0} \Delta \cdot \lambda_{C_1}(\Delta) = -\overset{-1}{K_1}(C_1) \tag{20.23}$$

* * *

$$\nabla I_2(\lambda_*, \Delta) = \begin{bmatrix} \partial_{\lambda_*} I_2(\lambda_*,\ \Delta) \\ \partial_\Delta I_2(\lambda_*,\ \Delta) \end{bmatrix} \tag{20.24}$$

$$\mathbf{t} \overset{\text{def.}}{=} \begin{bmatrix} -\partial_\Delta I_1(\lambda_*,\ \Delta) \\ \partial_{\lambda_*} I_1(\lambda_*,\ \Delta) \end{bmatrix} \tag{20.25}$$

$$\int_0^1 \int_0^1 \mathrm{d}\lambda \cdot \mathrm{d}\lambda' \cdot f(\lambda, \lambda') \cdot q(\lambda) \cdot q(\lambda') \overset{\text{id.}}{=} \int_0^1 \int_0^1 \mathrm{d}\lambda \cdot \mathrm{d}\lambda' \cdot \frac{[f(\lambda,\ \lambda') + f(\lambda',\ \lambda)]}{2} \cdot q(\lambda) \cdot q(\lambda') \tag{20.26}$$

$$\begin{aligned} -\frac{\text{sign}\ (\Delta) \cdot \mathbf{t}}{|\ \boldsymbol{t}\ |} \cdot \nabla I_2 = -\frac{|\ \Delta\ |}{|\ \boldsymbol{t}\ |} \cdot \int_0^1 \int_0^1 \mathrm{d}\lambda \cdot \mathrm{d}\lambda' \cdot (\lambda - \lambda_*)(\lambda' - \lambda) \cdot \partial_d \Phi[\lambda, \Delta(\lambda - \lambda_*)] \\ \cdot \partial_d \Phi[\lambda', \Delta(\lambda' - \lambda_*)] \overset{(20.26)}{=} \frac{|\ \Delta\ |}{2\ |\ \boldsymbol{t}\ |} \int_0^1 \int_0^1 \mathrm{d}\lambda \cdot \mathrm{d}\lambda' \\ \cdot (\lambda' - \lambda)^2 \cdot \partial_d \Phi[\lambda, \Delta(\lambda - \lambda_*)] \cdot \partial_d \Phi[\lambda', \Delta(\lambda' - \lambda_*)] \overset{(20.2)}{>} 0 \end{aligned} \tag{20.27}$$

* * *

$$\begin{aligned} \lim_{\Delta \to 0} I_2[\lambda_{C_1}(\Delta), \Delta] &= \lim_{\Delta \to 0} \int_0^1 \mathrm{d}\lambda \cdot \lambda \cdot \Phi\{\lambda, \Delta[\lambda - \lambda_{C_1}(\Delta)]\} \\ &= \int_0^1 \mathrm{d}\lambda \cdot \lambda \cdot \Phi\left\{\lambda, -\lim_{\Delta \to 0} \Delta \cdot \lambda_{C_1}(\Delta)\right\} \\ &\overset{(20.23)}{=} K_2\left[\overset{-1}{K_1}(C_1)\right] \overset{\text{def}}{=} \chi(C_1) \end{aligned} \tag{20.28}$$

* * *

$$\Delta \to \pm\infty : \tag{20.29}$$

$$\begin{aligned} I_2[\lambda_{C_1}(\Delta), \Delta] &= \int_0^1 \mathrm{d}\lambda \cdot \lambda \cdot \Phi\{\lambda, \Delta[\lambda - \lambda_{C_1}(\Delta)]\} \\ &= \int_0^1 \mathrm{d}\lambda \cdot \underbrace{[\lambda - \lambda_{C_1}(\Delta)] \cdot \Phi\{\lambda, \Delta[\lambda - \lambda_{C_1}(\Delta)]\}}_{\text{sign=sign}(\Delta)} + \underbrace{\lambda_{C_1}(\Delta) \cdot C_1}_{\text{bounded}} \end{aligned} \tag{20.30}$$

$$\left|\int_0^1 \mathrm{d}\lambda \cdot \underbrace{[\lambda - \lambda_{C_1}(\Delta)] \cdot \Phi\{\lambda, \Delta[\lambda - \lambda_{C_1}(\Delta)]\}}_{\mathrm{sign}=\mathrm{sign}(\Delta)}\right|$$
$$= \int_0^1 \mathrm{d}\lambda \cdot \left|[\lambda - \lambda_{C_1}(\Delta)] \cdot \Phi\{\lambda, \Delta[\lambda - \lambda_{C_1}(\Delta)]\}\right|$$
$$> \int_0^1 \mathrm{d}\lambda \cdot \left|[\lambda - \lambda_{C_1}(\Delta)] \cdot \check{\Phi}\{\Delta[\lambda - \lambda_{C_1}(\Delta)]\}\right| \overset{(20.4)}{\rightarrow} \infty \tag{20.31}$$

20.2 Existence and Uniqueness of the Solution

In Sect. 11.3.4 it was already shown that there is a spatially constant solution (11.71), if the constants C_1 and C_2 satisfy a corresponding relation (11.75). From the properties of the functions I_1 and I_2 it follows[3] that there is also a solution (11.76) – then spatially not constant – in all other cases.

Furthermore, it follows from these properties of the functions I_1 and I_2 that there is only one solution in each case. In the following this uniqueness of the solution is proved in a particularly simple way, by forming with two solutions d_1 and d_2 a positive definite expression, which can vanish only if both solutions agree. This positive definite expression must indeed vanish (20.35) due to the lateral boundary conditions being the same for both solutions (20.33) and due to the linearity of the two solutions (20.34). Therefore, the two solutions d_1 and d_2 must agree.

—

$$\sigma_i(\lambda) \overset{(11.61)}{=} \Phi[\lambda, d_i(\lambda)]; \quad i = 1,2 \tag{20.32}$$

$$0 \overset{(11.57),(11.58)}{=} \int_0^1 \mathrm{d}\lambda \cdot [\sigma_2(\lambda) - \sigma_1(\lambda)] = \int_0^1 \mathrm{d}\lambda \cdot \lambda \cdot [\sigma_2(\lambda) - \sigma_1(\lambda)] \tag{20.33}$$

$$0 \overset{(11.52)}{=} \partial_\lambda^2 d_i(\lambda); \quad i = 1,2 \tag{20.34}$$

$$0 = \int_0^1 \mathrm{d}\lambda \cdot \underbrace{[\sigma_2(\lambda) - \sigma_1(\lambda)] \cdot [d_2(\lambda) - d_1(\lambda)]}_{\overset{(20.2)}{\geq} 0} \tag{20.35}$$

[3] See No. 7, Sect. 20.1.

20.3 Examples

In the examples, the negative power $A^{-1/n}(\lambda)$ of the flow law parameter is supposed to change linearly in the vertical direction (20.36), where a temperature of ^{0} C with the corresponding value $A_{0^\circ\mathrm{C}}$ of the flow law parameter A *is* supposed to prevail at the bottom and a temperature T with the corresponding value A_T of the flow law parameter A *is* supposed to prevail at the surface.

If the criterion for constant spatial strain rates is not fulfilled (20.48), one obtains a determining equation for λ_* (20.50). Once one has determined λ_* from this, one can calculate Δ (20.51) and the horizontal strain rate d (20.49) is thus available.

If the criterion for spatially constant strain rates is satisfied (20.53), one obtains a spatially constant horizontal strain rate d_c (20.55) and (20.56). This case occurs only if the flow law parameter A_T at the surface has a value (20.54) determined by the criterion.

—

$$A^{-1/n}(\lambda) \overset{\text{def.}}{=} A_{0^\circ\mathrm{C}}^{-1/n} + \lambda \cdot \left[A_T^{-1/n} - A_{0^\circ\mathrm{C}}^{-1/n}\right] \tag{20.36}$$

$$\int_0^1 \mathrm{d}\lambda \cdot A^{-1/n}(\lambda) = \frac{1}{2} \cdot \left(A_{0^\circ\mathrm{C}}^{-1/n} + A_T^{-1/n}\right) \tag{20.37}$$

$$\int_0^1 \mathrm{d}\lambda \cdot \lambda \cdot A^{-1/n}(\lambda) = \frac{1}{6} \cdot \left(A_{0^\circ\mathrm{C}}^{-1/n} + 2A_T^{-1/n}\right) \tag{20.38}$$

$$\frac{\int_0^1 \mathrm{d}\lambda \cdot \lambda \cdot A^{-1/n}(\lambda)}{\int_0^1 \mathrm{d}\lambda \cdot A^{-1/n}(\lambda)} = \frac{1}{3} \cdot \frac{\left(1 + 2A_T^{-1/n}/A_{0^\circ\mathrm{C}}^{-1/n}\right)}{\left(1 + A_T^{-1/n}/A_{0^\circ\mathrm{C}}^{-1/n}\right)} \tag{20.39}$$

* * *

$$\mu \overset{\text{prec.}}{>} -1 \tag{20.40}$$

$$\partial_\lambda \mid \lambda - \lambda_* \mid \overset{\text{id.}}{=} \mathrm{sign}(\lambda - \lambda_*) \tag{20.41}$$

$$|\lambda - \lambda_*|^\mu \cdot \mathrm{sign}(\lambda - \lambda_*) \overset{\text{id.}}{=} \partial_\lambda \frac{|\lambda - \lambda_*|^{\mu+1}}{(\mu + 1)} \tag{20.42}$$

$$\int_0^1 \mathrm{d}\lambda \cdot |\lambda - \lambda_*|^\mu \cdot \mathrm{sign}(\lambda - \lambda_*) \overset{\text{id.}}{=} \frac{\left(|1 - \lambda_*|^{\mu+1} - |\lambda_*|^{\mu+1}\right)}{(\mu + 1)} \tag{20.43}$$

$$|\lambda - \lambda_*|^\mu \overset{\text{id.}}{=} \partial_\lambda \frac{\left[|\lambda - \lambda_*|^{\mu+1} \cdot \mathrm{sign}(\lambda - \lambda_*)\right]}{(\mu + 1)} \tag{20.44}$$

$$\int_0^1 \mathrm{d}\lambda \cdot |\lambda - \lambda_*|^\mu \overset{\text{id.}}{=} \frac{\left[|1 - \lambda_*|^{\mu+1} \cdot \mathrm{sign}(1 - \lambda_*) + |\lambda_*|^{\mu+1} \cdot \mathrm{sign}(\lambda_*)\right]}{(\mu + 1)} \tag{20.45}$$

$$\begin{aligned}
&\int_0^1 \mathrm{d}\lambda \cdot A^{-1/n}(\lambda) \cdot |\lambda - \lambda_*|^{1/n} \cdot \mathrm{sign}(\lambda - \lambda_*) \\
&= \frac{A_{0^\circ C}^{-1/n} \cdot \left[1 + \lambda_* \cdot \left(-1 + A_T^{-1/n}/A_{0^\circ C}^{-1/n}\right)\right]}{(1 + 1/n)} \cdot \left(|1 - \lambda_*|^{1+1/n} - |\lambda_*|^{1+1/n}\right) \\
&+ \frac{A_{0^\circ C}^{-1/n} \cdot \left(-1 + A_T^{-1/n}/A_{0^\circ C}^{-1/n}\right)}{(2 + 1/n)} \\
&\cdot \left[|1 - \lambda_*|^{2+1/n} \cdot \mathrm{sign}(1 - \lambda_*) + |\lambda_*|^{2+1/n} \cdot \mathrm{sign}(\lambda_*)\right]
\end{aligned} \tag{20.46}$$

$$\begin{aligned}
&\int_0^1 \mathrm{d}\lambda \cdot \lambda \cdot A^{-1/n}(\lambda) \cdot |\lambda - \lambda_*|^{1/n} \cdot \mathrm{sign}(\lambda - \lambda_*) \\
&= \frac{A_{0^\circ C}^{-1/n} \cdot \left(-1 + A_T^{-1/n}/A_{0^\circ C}^{-1/n}\right)}{(3 + 1/n)} \cdot \left(|1 - \lambda_*|^{3+1/n} - |\lambda_*|^{3+1/n}\right) \\
&+ \frac{A_{0^\circ C}^{-1/n} \cdot \left[1 + 2\lambda_* \cdot \left(-1 + A_T^{-1/n}/A_{0^\circ C}^{-1/n}\right)\right]}{(2 + 1/n)} \\
&\cdot \left[|1 - \lambda_*|^{2+1/n} \cdot \mathrm{sign}(1 - \lambda_*) + |\lambda_*|^{2+1/n} \cdot \mathrm{sign}(\lambda_*)\right] \\
&+ \frac{A_{0^\circ C}^{-1/n} \cdot \lambda_* \cdot \left[1 + \lambda_* \cdot \left(-1 + A_T^{-1/n}/A_{0^\circ C}^{-1/n}\right)\right]}{(1 + 1/n)} \cdot \left(|1 - \lambda_*|^{1+1/n} - |\lambda_*|^{1+1/n}\right)
\end{aligned} \tag{20.47}$$

* * *

$$C_1 \overset{\text{prec.}}{\neq} 0; \quad \frac{1}{3} \cdot \frac{\left(1 + 2A_T^{-1/n}/A_{0^\circ C}^{-1/n}\right)}{\left(1 + A_T^{-1/n}/A_{0^\circ C}^{-1/n}\right)} \overset{(11.96),(20.39)}{\neq} \frac{C_2}{C_1} \tag{20.48}$$

$$d(\lambda) = \Delta \cdot (\lambda - \lambda_*) \tag{20.49}$$

$$\begin{aligned}
&\frac{\left[|1-\lambda_*|^{3+1/n} - |\lambda_*|^{3+1/n}\right]}{(3+1/n)} \cdot \left[\left(A_T/A_{0^\circ C}\right)^{-1/n} - 1\right] \\
&+ \frac{\left[|1-\lambda_*|^{2+1/n} \cdot \mathrm{sign}(1-\lambda_*) + |\lambda_*|^{2+1/n} \cdot \mathrm{sign}(\lambda_*)\right]}{(2+1/n)} \\
&\cdot \left\{1 + [2\lambda_* - (C_2/C_1)]\left[\left(A_T/A_{0^\circ C}\right)^{-1/n} - 1\right]\right\} \\
&+ \frac{\left[|1-\lambda_*|^{1+1/n} - |\lambda_*|^{1+1/n}\right]}{(1+1/n)} \cdot \left\{1 + \lambda_*\left[\left(A_T/A_{0^\circ C}\right)^{-1/n} - 1\right]\right\}[\lambda_* - (C_2/C_1)] = 0
\end{aligned} \tag{20.50}$$

[→ (11.100); (20.46); (20.47)]

$$|\Delta|^{1/n}\mathrm{sign}(\Delta) = C_1\sqrt{3}^{(1-1/n)}\left\{\underbrace{\int_0^1 d\lambda \cdot A^{-1/n}(\lambda) \cdot |\lambda - \lambda_*|^{1/n} \cdot \mathrm{sign}(\lambda - \lambda_*)}_{(20.46)}\right\}^{-1} \tag{20.51}$$

[→ (11.101)]

* * *

$$C_1 \overset{\text{prec.}}{\neq} 0 \tag{20.52}$$

$$\frac{1}{3} \cdot \frac{\left(1 + 2A_T^{-1/n}/A_{0^\circ \mathrm{C}}^{-1/n}\right)}{\left(1 + A_T^{-1/n}/A_{0^\circ \mathrm{C}}^{-1/n}\right)} \overset{(11.102),(20.39)}{=} \frac{C_2}{C_1} \tag{20.53}$$

$$\left(\frac{A_T}{A_{0^\circ \mathrm{C}}}\right)^{-1/n} = \frac{(-1 + 3C_2/C_1)}{(2 - 3C_2/C_1)} \tag{20.54}$$

$$d(\lambda) = d_c \tag{20.55}$$

$$|d_c|^{1/n} \cdot \mathrm{sign}(d_c) \overset{(11.105),(20.37)}{=} C_1 \cdot \sqrt{3}^{(1-1/n)} \cdot 2\left(A_{0^\circ \mathrm{C}}^{-1/n} + A_T^{-1/n}\right)^{-1} \tag{20.56}$$

20.4 The Constants C_1 and C_2

20.4.1 Spatially Non-constant Densities of Ice and Water

It is calculated with the dimensionless vertical coordinate λ (20.59), which disappears at the bottom of the table iceberg and has the value 1 at its surface. The thickness of the table iceberg is denoted by h (20.57) and the thickness of the water body replaced by the tabular iceberg by $\widetilde{h}$ (20.58). All quantities can be expressed uniformly by multiple integrals by using the ∂_λ^{-1} integral operator (20.63)[4] and continuing all functions (20.62) that occur, such as the densities ρ (20.60) and $\widetilde{\rho}$ (20.61) of ice and water, upward through the value zero beyond their original domain of definition.

Thus Archimedes' principle (20.70) and the constants C_1 (20.71) and C_2 (20.72) can be expressed by single or multiple integrals of the ice-water density difference.

—

$$h = z_0 - z_1 \tag{20.57}$$

$$\widetilde{h} = -z_1 \tag{20.58}$$

$$\lambda = \frac{z - z_1}{z_0 - z_1} = \frac{z + \widetilde{h}}{h} \tag{20.59}$$

* * *

$$\rho(\lambda) \overset{\text{def.}}{=} 0; \quad \lambda \overset{\text{prec.}}{>} 1 \tag{20.60}$$

$$\widetilde{\rho}(\lambda) \overset{\text{def.}}{=} 0; \quad \lambda \overset{\text{prec.}}{>} \widetilde{h}/h \tag{20.61}$$

$$\psi(\lambda) \overset{\text{prec.}}{=} 0; \quad \lambda \overset{\text{prec.}}{>} 1 \tag{20.62}$$

$$\left(\partial_\lambda^{-1}\psi\right)(\lambda) \overset{\text{def.}}{=} -\int_\lambda^\infty d\lambda' \cdot \psi(\lambda') \tag{20.63}$$

$$\partial_\lambda^{-1}(\lambda \cdot \psi) \overset{\text{id.}}{=} \lambda \cdot \partial_\lambda^{-1}\psi - \partial_\lambda^{-2}\psi \tag{20.64}$$

[4] See Sect. 3.1.

$$\left[\partial_\lambda^{-1}(\lambda \cdot \psi)\right]_{\lambda=0} \overset{\text{id.}}{=} -\left[\partial_\lambda^{-2}\psi\right]_{\lambda=0} \tag{20.65}$$

* * *

$$p(\lambda) = -gh \cdot \partial_\lambda^{-1}\rho \tag{20.66}$$

$$\widetilde{p}(\lambda) = -gh \cdot \partial_\lambda^{-1}\widetilde{\rho} \tag{20.67}$$

$$\partial_\lambda^{-n} p = -gh \cdot \partial_\lambda^{-n-1}\rho; \quad n = 0,1,\ldots \tag{20.68}$$

$$\partial_\lambda^{-n} \tilde{p} = -gh \cdot \partial_\lambda^{-n-1}\tilde{\rho}; \quad n = 0,1,\ldots \tag{20.69}$$

* * *

$$0 = \left[\partial_\lambda^{-1}(\rho - \tilde{\rho})\right]_{\lambda=0} \tag{20.70}$$

$$C_1 \overset{(11.46)}{=} -\frac{1}{3} \cdot \left[\partial_\lambda^{-1}(p - \widetilde{p})\right]_{\lambda=0} = \frac{gh}{3} \cdot \left[\partial_\lambda^{-2}(\rho - \widetilde{\rho})\right]_{\lambda=0} \tag{20.71}$$

$$C_2 \overset{(11.47)}{=} -\frac{1}{3} \cdot \left\{\partial_\lambda^{-1}[\lambda \cdot (p - \widetilde{p})]\right\}_{\lambda=0} \overset{(20.65)}{=} -\frac{gh}{3} \cdot \left[\partial_\lambda^{-3}(\rho - \widetilde{\rho})\right]_{\lambda=0} \tag{20.72}$$

$$\frac{C_2}{C_1} = -\left[\frac{\partial_\lambda^{-3}(\rho - \tilde{\rho})}{\partial_\lambda^{-2}(\rho - \tilde{\rho})}\right]_{\lambda=0} \tag{20.73}$$

20.4.2 Spatially Constant Densities of Ice and Water

If the densities in the table iceberg and in the water body have spatially constant values ρ_c (20.74) and $\widetilde{\rho}_c$ (20.75), respectively, Archimedes' principle (20.78) and the constants C_1 (20.79) and C_2 (20.80) can be expressed by these values.

—

$$\rho(\lambda) = \theta(1 - \lambda) \cdot \rho_c \tag{20.74}$$

$$\widetilde{\rho}(\lambda) = \theta\left(\frac{\widetilde{h}}{h} - \lambda\right) \cdot \widetilde{\rho}_c = \theta\left(\frac{\rho_c}{\widetilde{\rho}_c} - \lambda\right) \cdot \widetilde{\rho}_c \tag{20.75}$$

$$\partial_\lambda^{-n} \rho = \theta(1 - \lambda) \cdot \frac{(-1)^n}{n!} \cdot (1 - \lambda)^n \cdot \rho_c; \quad n = 0,1,\ldots \tag{20.76}$$

$$\partial_\lambda^{-n} \widetilde{\rho} = \theta\left(\frac{\rho_c}{\widetilde{\rho}_c} - \lambda\right) \cdot \frac{(-1)^n}{n!} \cdot \left(\frac{\rho_c}{\widetilde{\rho}_c} - \lambda\right)^n \cdot \widetilde{\rho}_c; \quad n = 0,1,\ldots \tag{20.77}$$

* * *

$$\frac{\rho_c}{\widetilde{\rho}_c} = \frac{\widetilde{h}}{h} \tag{20.78}$$

$$C_1 = \frac{gh\rho_c}{6} \cdot \left(1 - \frac{\rho_c}{\widetilde{\rho}_c}\right) \tag{20.79}$$

$$C_2 = \frac{gh\rho_c}{18} \cdot \left(1 - \frac{\rho_c^2}{\widetilde{\rho}_c^2}\right) \tag{20.80}$$

$$\frac{C_2}{C_1} = \frac{1}{3} \cdot \left(1 + \frac{\rho_c}{\widetilde{\rho}_c}\right) \tag{20.81}$$

Appendix A: Explanation and List of Symbols

For Cartesian coordinates, numbered and unnumbered designations are used synonymously with coordinates $x = x_1$, $y = x_2$, $z = x_3$, $\partial_x = \partial_1$ etc. for partial derivatives, $u_x = u_1$ etc. for vector components and $H_{xx} = H_{11}$ etc. for tensor components. The summation convention applies, according to which summation is to be performed over index variables occurring in pairs.

Vectors and tensors are denoted by bold Latin letters, each of which can have a double meaning. They may denote both the vector or tensor in question and the matrix of its Cartesian components with respect to a defined coordinate system. The symbol $\mathbf{r}$ can occur in three related meanings: It may symbolize the location vector from the coordinate origin to the particular point under consideration, or the column matrix of its Cartesian components x, y, z, or the point itself. Some symbols, such as χ, are used in several unrelated meanings and are listed in the following list accordingly often.

Products written with or without a dot are either scalar products between column matrices or general matrix products defined according to the rules of matrix calculus, such as the scalar $|\mathbf{n}|^2 = \mathbf{nn} = \mathbf{n} \cdot \mathbf{n} = \mathbf{n}^{\mathrm{T}}\mathbf{n} = \mathbf{n}^{\mathrm{T}} \cdot \mathbf{n} = 1$ of a unit vector $\mathbf{n}$ or the matrix $\mathbf{nn}^{\mathrm{T}} = \mathbf{n} \cdot \mathbf{n}^{\mathrm{T}}$ of the projector on the direction defined by $\mathbf{n}$. The transposition of a square matrix $\mathbf{H}$ and its symmetric or antisymmetric part are denoted by $\mathbf{H}^{\mathrm{T}}$ resp. $\mathbf{H}_+$ resp. $\mathbf{H}_-$. Slashes through bold symbols denote mappings between vectors and antisymmetric tensors: $\not{\mathbf{u}}$ denotes the antisymmetric tensor associated with a vector $\mathbf{u}$ and $\not{\mathbf{H}}$ denotes the vector associated with the antisymmetric part $\mathbf{H}_-$ of a tensor $\mathbf{H}$. Differential operators are sometimes used unconventionally, such as the ∇-operator when it acts to the left in the divergence formation of a tensor field or the gradient formation of a vector field, instead of to the right as usual. In such unconventional cases, the functions on which a differential operator acts are indicated by a vertical arrow.

A slash above a function, such as $\overline{\rho}$, denotes a smooth (infinitely often differentiable) function. For a function f defined on a domain of definition Ω_{def}, the symbols $[f]_{z=z_0(x,y)}$ or $[f]_{z=z_0}$ or $[f]_{z_0}$ denote a function also defined on Ω_{def} independently of z, whose values at a point (x, y, z) are given by the values of the function f at the point $(x, y, z_0(x, y))$. If a function

P. Halfar, *Stresses in glaciers*, https://doi.org/10.1007/978-3-662-66024-9

f defined on a domain of definition Ω_{def} is considered only on a surface Σ ($\Sigma \subset \Omega_{def}$), then this function given on its restricted domain of definition Σ is denoted by $[f]_\Sigma$.

A, A_T	Flow law parameters, at temperature T (Sects. 11.3.5, 11.3.6, and 20.3)
$\mathbf{A}$	Stress function from which its successors, $\mathbf{B}$, $\mathbf{C}$ and $\mathbf{T}$ descend (Sect. 6.1)
$\mathbf{A}_*$	Stress function with successor $\mathbf{T}_*$, produces the boundary stresses $\mathbf{t}$ on the surface Σ and no forces and torques on the surfaces Λ_μ ($\mu \neq 0$) (Sects. 7.3 and 16.1)
$\mathbf{A}_{**}$	Stress function with successor $\mathbf{T}_{**}$, generates on Σ no boundary stresses and on the surfaces Λ_μ the forces $\mathbf{F}_\mu$ and torques $\mathbf{G}_\mu$ ($\mu \neq 0$) (Sects. 7.3 and 16.2)
$\mathbf{A}_0$	Stress function, vanishes together with its first derivatives on Σ (Sects. 7.3, 9.1.1, and 14.3)
$\mathbf{A}_\Sigma$	Boundary value of $\mathbf{A}$ on Σ (Sect. 7.1)
$\mathbf{A}^\bullet$	Redundancy function, stress function with successor $\mathbf{T}^\bullet = \mathbf{0}$ (Sects. 6.2.1 and 14.1)
$\mathbf{A}_\mu$	Stress function, generates on Σ no boundary stresses, on the surface Λ_μ the force $\mathbf{F}_\mu$ and torque $\mathbf{G}_\mu$ ($\mu \neq 0$), on the surface Λ_0 the force $-\mathbf{F}_\mu$ and torque $-\mathbf{G}_\mu$ and on the other surfaces Λ_ν no boundary stresses and therefore no forces and torques (Sect. 16.2)
dA	Area element (Chap. 2, Sect. 15.1)
$\mathbf{a}$	Point on the base of a table iceberg or location vector of this point (Sect. 11.3.2).
$\mathbf{B}$	Matrix field, successor of $\mathbf{A}$ (Sect. 6.1)
$\mathbf{B}_\Sigma$	Boundary value of $\mathbf{B}$ on Σ (Sect. 7.1)
$\mathbf{B}^\bullet$	Matrix field, successor of $\mathbf{A}^\bullet$ (Sects. 6.2.1 and 14.1)
$\mathbf{B}_{**}$	Matrix field, successor of $\mathbf{A}_{**}$ (Sect. 16.2)
$\mathbf{B}_\mu$	Matrix field, successor of $\mathbf{A}_\mu$ (Sect. 16.2)
$\mathbf{b}$	Point on the base of a table iceberg or location vector of this point (Sect. 11.3.2).
$\mathbf{b}_1, \mathbf{b}_2\ \mathbf{b}_n$	Column matrix fields, evolution coefficients of $\mathbf{B}_\Sigma$ (Sect. 15.4)
$\mathbf{C}$	Matrix field, successor of $\mathbf{A}$ (Sect. 6.1)
$\mathbf{C}_\Sigma$	Boundary value of $\mathbf{C}$ on Σ (Sect. 7.1)
$\mathbf{C}^\bullet$	Matrix field, successor of $\mathbf{A}^\bullet$ (Sects. 6.2.1 and 14.1)
$\mathbf{C}_{**}$	Matrix field, successor of $\mathbf{A}_{**}$ (Sect. 16.2)
$\mathbf{C}_\mu$	Matrix field, successor of $\mathbf{A}_\mu$ (Sect. 16.2)
C_1, C_2	Constant in the lateral boundary conditions for table icebergs (Sect. 11.3.4, Chap. 20)
$\mathbf{c}_\omega$	Center-of-mass vector of the ice mass in area ω (Chap. 2).
$\mathbf{c}$	Center-of-mass vector of an iceberg (Sect. 11.2)
$\tilde{\mathbf{c}}$	Center-of-mass vector of a water body (Sect. 11.2)
$\mathbf{c}_1, \mathbf{c}_2\ \mathbf{c}_n$	Column matrix fields, evolution coefficients of $\mathbf{C}_\Sigma$ (Sect. 15.4)
$\mathbf{D}$	Tensor field of strain rates (Sect. 11.3.3)
D	Invariant of $\mathbf{D}$ (Sect. 11.3.5)
d	Horizontal strain rate (Sect. 11.3.3)
d_c	Spatially constant horizontal strain rate (Sect. 11.3.3)
div	Divergence operator, applied to vector or tensor fields (see (13.8))

(continued)

$\mathbf{e}_1, \mathbf{e}_2\ \mathbf{e}_3$	Orthonormal basis (Chap. 12)
$\mathbf{F}_{\mathbf{a},\ \mathbf{b}}$	Force on a vertical intersection in the table iceberg (Sect. 11.3.2)
$\tilde{\mathbf{F}}_{\mathbf{a},\mathbf{b}}$	Hydrostatic force on a vertical intersection in the table iceberg (Sect. 11.3.2)
$\mathbf{F}_{\Sigma}, \mathbf{F}_{\Sigma}[\mathbf{T}]$	Force of **T** onto Σ (Sect. 7.1)
$\mathbf{F}_{\nu}, \mathbf{F}_{\nu}[\mathbf{T}]$	Force of **T** onto Λ_{ν} (Sect. 7.1)
$\mathbf{F}_{\nu}[\mathbf{S}]$	Force of **S** onto Λ_{ν} (Sect. 8.1)
$\mathbf{F}_{\nu}[\mathbf{S}_{\mathrm{bal}}]$	Force of $\mathbf{S}_{\mathrm{bal}}$ onto Λ_{ν} (Sect. 8.1)
$\mathbf{f}_1, \mathbf{f}_2$	Tangential basis vectors on Σ (Sect. 15.1)
$\mathbf{f}^1, \mathbf{f}^2$	Dual tangential basis vectors on Σ (Sect. 15.1)
$\mathbf{G}_{\mathbf{a},\ \mathbf{b}}$	Torque on a vertical intersection in the table iceberg (Sect. 11.3.2)
$\tilde{\mathbf{G}}_{\mathbf{a},\mathbf{b}}$	Hydrostatic torque on a vertical intersection in the table iceberg (Sect. 11.3.2)
$\mathbf{G}_{\Sigma}, \mathbf{G}_{\Sigma}[\mathbf{T}]$	Torque of Tonto Σ (Sect. 7.1)
$\mathbf{G}_{\nu}, \mathbf{G}_{\nu}[\mathbf{T}]$	Torque of **T** onto Λ_{ν} (Sect. 7.1)
$\mathbf{G}_{\nu}[\mathbf{S}]$	Torque of **S** onto Λ_{ν} (Sect. 8.1)
$\mathbf{G}_{\nu}[\mathbf{S}_{\mathrm{bal}}]$	Torque of $\mathbf{S}_{\mathrm{bal}}$ onto Λ_{ν} (Sect. 8.1)
G	Function for solving the hyperbolic differential equation (Sect. 3.2)
g	Amount of acceleration due to gravity (10.95)
$\mathbf{g}$	Acceleration due to gravity (Chap. 2)
grad	Gradient operator, applied to scalar or vector fields (see (13.5))
h	Ice thickness of a table iceberg (Sects. 11.3.2 and 20.4.1)
$\tilde{h}$	Immersion depth of a table iceberg (Sects. 11.3.2 and 20.4.1)
$\mathbf{h}$	Horizontal normal vector (Sect. 10.3.3)
I_1	Function in the lateral boundary conditions for tabular icebergs (Sects. 11.3.4 and 20.1)
I_2	(Sects. 11.3.4 and 20.1)
K_1	(Sects. 11.3.4 and 20.1)
K_2	(Sects. 11.3.4 and 20.1)
$\mathbf{k}$	Cone vector of the model cone generated by all integration cones (Sect. 3.2)
$\mathbf{k}$	Column matrix field on Σ, occurs in the calculation of the normal derivative $\partial_n \mathbf{A}$ (Sect. 15.3).
$\mathcal{L}$	Matrix differential operator, occurs in the calculation of the general solution of the balance and boundary conditions from three independent stress components (Sects. 3.4.2, (17.4)–(17.64))
$\mathcal{L}_{\mathrm{adj}}$	Adjoint to $\mathcal{L}$ (Sect. 3.4.3)
$\det(\mathcal{L})$	Determinant of $\mathcal{L}$, differential operator (Sect. 3.4.3)
$\mathcal{L}^{-1}$	Inverse of $\mathcal{L}$, matrix differential integral operator for computing the general solution of the balance and boundary conditions from three independent stress components (Sect. 3.4.3, (17.5)–(17.67))
$[\det(\mathcal{L})]^{-1}$	Integral operator (Sect. 3.4.3)
$\mathbf{M}_{\omega}$	Moment of the ice mass in the range ω (Chap. 2)
$\mathbf{M}_z(\Gamma)$	Projection moment, moment of ice mass in z-direction of Γ (Sect. 4.4)
m	Iceberg mass (Sect. 11.2)

(continued)

$\tilde{m}$	Water mass (Sect. 11.2)
m_ω	Ice mass in the range ω (Chap. 2)
$m_z(\Gamma)$	Projection mass, ice mass in the z-direction of Γ (Sect. 4.4)
$\mathbf{n}$	Unit vector, normal vector to Σ (Chaps. 2 and 12, Sect. 15.1)
$\check{\mathbf{n}}$	Normal vector at the horizontal surface of a stagnant glacier (see (10.84))
$\mathbf{n}_0$, $\mathbf{l}_0$ $\mathbf{m}_0$	Orthonormal base at the glacier surface aligned with the steepest slope and the contour line (Sect. 10.3.3)
$\mathbf{n}_1$, $\mathbf{l}_1$ $\mathbf{m}_1$	Orthonormal base at the glacier bottom aligned with the steepest slope and the contour line (Sect. 10.3.3)
$\mathbf{P}$	Projector onto the direction of **n** (Chap. 12, Sect. 15.1)
p	Gravity pressure of the ice (Sects. 10.3.3 and 11.1)
$\tilde{p}$	Hydrostatic pressure (Chaps. 2 and 11)
$\check{p}$	Gravity pressure of ice in a stagnant glacier (Sect. 10.3.1)
p_1	Gravity pressure of the ice at the glacier bed (Sect. 10.3.3)
$\mathbf{Q}$	Projector onto the tangential plane of Σ (Chap. 12, Sect. 15.1)
q_1, q_2, q_3	Functions in the distributional formulation of the hyperbolic differential equation (Sect. 18.6, Chap. 19)
R	Distance from an axis of rotation (Sect. 16.2)
R_x, R_y	Radii of curvature of the glacier surface in *x* and *y* direction (Sect. 10.3.3)
$\mathbf{r}$	Location vector (Chaps. 2 and 12)
rot	Rotation operator, applied to vector or tensor fields (see (13.6))
$\mathbf{r}_\Sigma$	Location vector of points on Σ (Sect. 15.1)
$d\mathbf{r}$	Vector path element
$\mathbf{S}$	Stress tensor, stress tensor field (Chap. 2)
$\mathbf{S}'$	Deviatoric stress tensor, d. stress tensor field (see (11.38) and (11.79))
$\mathbf{S}_*$	Stress tensor field with three vanishing components (Sects. 8.2.1 and 9.1.2, (17.9)–(17.71))
$\mathbf{S}'_*$	Deviatoric stress tensor field, arises from $\mathbf{S}_*$ (see (8.6))
S'	Invariant of $\mathbf{S}'$ (Sect. 11.3.5)
$\hat{\mathbf{S}}$	Quasi-stagnant stress tensor field (Sect. 10.3.3)
$\check{\mathbf{S}}$	Stress tensor field in a stagnant glacier (Sect. 10.3.1)
$\tilde{\mathbf{S}}$	Hydrostatic tensor (Chap. 11)
$\mathbf{S}_a, \ldots \mathbf{S}_h$	Stress tensor fields with three vanishing components (see (17.9)–(17.71))
$\mathbf{S}_b$	Stress tensor field with vanishing *xx*, *yy* and *xy* components (Sects. 5.1 and 10.1.1, see (17.17))
$\mathbf{S}_{\text{bal}}$	Stress tensor field, special solution of balance conditions (Chap. 2)
$\mathbf{S}_e$	Stress tensor field with vanishing non-diagonal components (Sect. 5.2, (17.41), (10.27))
$\mathbf{S}_f$	Stress tensor field with vanishing deviatoric *xx*, *yy* and *xy* components (Sect. 10. 1.3, (17.49))
$\mathbf{s}$	Boundary stress on Σ (Chap. 2)
$\mathbf{T}$	Weightless stress tensor field, derived from **A** (Chap. 2, Sect. 6.1)

(continued)

$\mathbf{T}_*$	Weightless stress tensor field, is derived from $\mathbf{A}_*$ (Sects. 7.3 and 16.1)
$\mathbf{T}_{**}$	Weightless stress tensor field, derives from $\mathbf{A}_{**}$ (Sects. 7.3, 9.1.1, 10.2, and 16.2)
$\mathbf{T}_0$	Weightless stress tensor field, derived from $\mathbf{A}_0$ (Sects. 7.3 and 9.1.1)
$\mathbf{T}_\mu$	Weightless stress tensor field, derived from $\mathbf{A}_\mu$ (Sect. 16.2, (16.13))
$\mathbf{T}^\bullet$	Vanishing stress tensor field, derived from $\mathbf{A}^\bullet$ (Sect. 6.2.1)
$\mathbf{t}$	Boundary stresses of $\mathbf{T}$ on Σ (Chap. 2)
$\mathbf{v}$	Ice flow velocity (Sect. 11.3.3)
dV	Volume element (Chap. 2)
$\mathbf{w}_\mu$	Vector field for calculating $\mathbf{A}_\mu$ (Sect. 16.2)
$x_0(y, z)$	Function for displaying the free glacier surface (Sect. 18.1)
x', y', x_1', x_2'	Curvilinear surface coordinates on Σ (Sect. 15.1)
$y_0(x, z)$	Function for displaying the free glacier surface (Sect. 18.1)
$z_0(x, y)$	Function for displaying the free glacier surface ((15.16), Sects. 11.1 and 18.1)
$z_1(x, y)$	Function for displaying the glacier bed (Sects. 10.3.3 and 11.1)
α_x, α_y	Surface slope angle in x or y direction (Sect. 10.3.3)
$\alpha_{\max}$	Angle of surface slope in the direction of maximum slope (Sect. 10.3.3)
$\beta_{\max}$	Angle of bottom slope in the direction of maximum slope (Sect. 10.3.3)
Γ	Oriented surface in the glacier (Sect. 4.1)
$\partial\Gamma$	Oriented boundary curve of Γ (Sect. 4.1)
$\varDelta$	Difference of strain rates at the top and bottom of a table iceberg (Sect. 11.3.3)
δ	Dirac's delta function (Sects. 3.1 and 18.2)
δ'	Derivative of the delta function (Sect. 18.2)
δ''	Second derivative of the delta function (Sect. 18.2)
δ_{ij}	Kronecker symbol (Chap. 12)
ε	Function (Sect. 10.3.3)
ε_{ijk}	Antisymmetric tensor components (Chap. 12)
θ	Heaviside or step function (Sects. 3.1 and 18.2)
K_x, K_y, K_z	Integration cone of ∂_x^{-1} etc. (Sect. 8.2.2)
K_{xy}, K_{yz}, K_{xz}	Integration cone of $\partial_x^{-1}\partial_y^{-1}$ etc. (Sect. 8.2.2)
K_{xyz}	Integration cone of $\partial_x^{-1}\partial_y^{-1}\partial_z^{-1}$ (Sect. 8.2.2)
$K_z^{\odot}$	Integration cone of $\square_z^{-1}$ (Sect. 8.2.2)
K'_{xz}	Integration cone of $\left(\partial_x^2 - 2\partial_z^2\right)^{-1}$ (Sect. 8.2.2)
K''_{xz}	Integration cone of $\left(\partial_x^2 - 2\partial_z^2\right)^{-1}$ (Sect. 8.2.2)
K'_{yxz}	Integration cone of $\partial_y^{-1}\left(\partial_x^2 - 2\partial_z^2\right)^{-1}$ (Sect. 8.2.2)
K''_{yxz}	Integration cone of $\partial_y^{-1}\left(\partial_x^2 - 2\partial_z^2\right)^{-1}$ (Sect. 8.2.2)
Λ	Boundary surface of unknown boundary stresses, rests on $\partial\Omega$ (Sect. 7.1)
Λ_ν ($\nu = 0, 1, \ldots n$)	Separate, connected parts of Λ (Sect. 7.1)

(continued)

$\partial\Lambda_\nu$	Closed boundary curve of Λ_ν (Sect. 7.1)
λ	Dimensionless vertical coordinate in a table iceberg (Sect. 11.3.2)
λ_*	Dimensionless vertical coordinate of vanishing strain rates in a table iceberg (Sect. 11.3.3)
$\lambda_{C_1}(\Delta)$	Level line of function I_1 at value C_1 as a function of Δ (Sect. 20.1)
ρ	Ice density Chap. 2)
$\breve{\rho}$	Horizontally homogeneous ice density in a stagnant glacier (Sect. 10.3.1)
ρ_c	Spatially constant ice density (Sects. 10.3.3, 11.3.6, and 20.4.2)
$\tilde{\rho}$	Water density (Chap. 11, Sect. 20.4.1)
$\tilde{\rho}_c$	Spatially constant water density (Sects. 11.3.6 and 20.4.2)
Σ	Boundary surface on which the boundary stresses are specified (Chap. 2, Sect. 18.1)
$\breve{\Sigma}$	Horizontal free surface of a stagnant glacier (Sect. 10.3.1)
Σ_0	Top of the glacier (Sect. 11.1)
Σ_1	Glacier bed, underside (Sects. 10.3.3 and 11.1)
$\Sigma_\perp$	Vertical boundary (Sect. 11.1)
σ	Deviatoric longitudinal stress (Sect. 11.3.2)
τ	Test function (Sect. 3.5.2)
Φ	General flow law function in the horizontally isotropic-homogeneous case (Sect. 11.3.4)
$\breve{\Phi}$	Function characterising a property of Φ (Sect. 11.3.4)
ϕ	Angular coordinate with respect to an axis of rotation (Sect. 16.2)
χ	Distribution (Sect. 3.5.2)
χ	Function for the analysis of strain rates in a table iceberg (Sects. 11.3.4 and 20.1)
χ	Function, solution of the hyperbolic differential equation with boundary conditions (Chap. 19)
χ	Interpolation function, used to construct a non-singular vector field $\mathbf{w}_\mu$ (Sect. 16.2)
ψ	Smooth function (Sects. 18.3, 18.4, 18.5, and 18.6)
Ω	Glacier area under consideration (Chap. 2, Sect. 18.1)
$\partial\Omega$	Closed boundary of Ω (Chap. 2)
Ω_{def}	Domain of definition, contains Ω (Sects. 3.4.1 and 18.1)
Ω_{ext}	External part of Ω_{def} (Sects. 3.4.1 and 18.1)
ω	Part of Ω (Chap. 2)
$\partial\omega$	Closed boundary of ω (Chap. 2)
$\omega_z(\Gamma)$	Projection shadow of Γ cast in the z-direction (Sect. 4.1)
∇	Nabla operator (see (13.1))
∇	Rotation operator (see (13.2), Sect. 15.2)
∂_n	Derivation in the direction of the surface normal $\mathbf{n}$ (see (13.3), Sect. 15.2)
∂'_1, ∂'_2	Differential operators on Σ (Sect. 15.2)
$\partial_x^{-1}, \partial_y^{-1}$ ∂_z^{-1}	Integral operators (Sect. 3.1)

(continued)

$(\mathbf{a}\nabla)^{-1}$, $(\mathbf{a}\nabla)^{-1*}$	Integral operators (Sects. 3.2 and 3.5.2)
$\Box_z$	Hyperbolic differential operator (Sect. 3.2)
$\Box_z^{-1}$, $\Box_z^{-1*}$	Integral operators (Sects. 3.2 and 3.5.2)

References

1. Goldstein, H.: Klassische Mechanik. Akademische Verlagsgesellschaft, Wiesbaden (1978)
2. Gurtin, M.E.: The linear theory of elasticity. In: Truesdell, C. (ed.) Festkörpermechanik II, pp. 1–296. In: Flügge, S. (ed.) Handbuch der Physik, Vol. VIa/2. Springer, Berlin (1972)
3. Nye, J.F.: A comparison between the theoretical and the measured long profile of the Unteraar Glacier. J. Glaciol. **2**, 103–107 (1952)
4. Paterson, W.S.B.: The Physics of Glaciers. Elsevier, Oxford (1994)
5. Serrin, J.: Mathematical principles of fluid mechanics. In: Truesdell, C. (ed.) Strömungsmechanik I, pp. 125–263. In Flügge, S. (ed.) Handbuch der Physik, Vol. VIII/1. Springer, Berlin (1959)

P. Halfar, *Stresses in glaciers*, https://doi.org/10.1007/978-3-662-66024-9

Printed in the United States
by Baker & Taylor Publisher Services